AF389161

Cutting-Edge Technologies for Renewable Energy Production and Storage

Cutting-Edge Technologies for Renewable Energy Production and Storage

Special Issue Editor

Matteo Prussi

MDPI • Basel • Beijing • Wuhan • Barcelona • Belgrade • Manchester • Tokyo • Cluj • Tianjin

Special Issue Editor
Matteo Prussi
European Commission Joint
Research Centre (JRC)
Italy

Editorial Office
MDPI
St. Alban-Anlage 66
4052 Basel, Switzerland

This is a reprint of articles from the Special Issue published online in the open access journal *Applied Sciences* (ISSN 2076-3417) (available at: https://www.mdpi.com/journal/applsci/special_issues/Renewable_Energy_Production_Storage).

For citation purposes, cite each article independently as indicated on the article page online and as indicated below:

LastName, A.A.; LastName, B.B.; LastName, C.C. Article Title. *Journal Name* **Year**, *Article Number*, Page Range.

ISBN 978-3-03936-000-0 (Pbk)
ISBN 978-3-03936-001-7 (PDF)

Disclaimer

The views expressed here are purely those of the authors and may not, under any circumstances, be regarded as an official position of the European Commission.

Contents

About the Special Issue Editor

Matteo Prussi is an industrial engineer, with a scientific background in renewable energy conversion technologies. He has been working in biofuels sector for more than 10 years.

He is currently employed at the Sustainable Transport Unit of the EC JRC. His activity focuses on the assessment of the environmental impacts of the various modes of transport and on the potential for alternative sustainable fuels, in particular for aviation sector. In the aviation sector, he has been nominated by EC as co-leader of the CORE-LCA subgroup of ICAO-Alternative Fuel Task Group (FTG).

He was a research fellow in the Energy Department at the University of Florence. Between 2012 and 2017, he covered the role of Director of RE-CORD renewable energy consortium. Part of his research activities, supported by European projects, focused on alternative technologies for biofuels production and second generation bioethanol, with appraisals of their potential for aviation applications.

Editorial

Special Issue on Cutting-Edge Technologies for Renewable Energy Production and Storage

Matteo Prussi

European Commission Joint Research Centre (JRC), Directorate C—Energy, Transport and Climate—C.2, ISPRA, Via E. Fermi 2749, TP 023 I, 27027 Ispra, Italy; matteo.prussi@ec.europa.eu

Received: 16 January 2020; Accepted: 20 January 2020; Published: 24 January 2020

1. Introduction

Anthropogenic greenhouse gas emissions are dramatically influencing the environment, and research is strongly committed in proposing alternatives, mainly based on renewable energy sources [1]. Low-greenhouse gas (GHG) electricity production from renewables is well established, but issues of grid balancing are limiting its development. Energy storage is a key topic for the further deployment of renewable energy production [2]. Besides battery and other types of electrical storage, electrofuels and bioderived fuels may offer suitable alternatives in some specific scenarios [3,4]. This Special Issue welcomed contributions on energy conversion technologies and use, energy storage, technologies integration, e-fuels, and pilot and large-scale applications.

2. Technologies for Production and Storage: Enabling Larger Renewable Energy Uptake

In light of the above, this Special Issue collected the latest research papers on relevant topics about energy production and use. As the predictability of the output from a power production plant is a key element for a stable, reliable energy infrastructure [5], the Special Issue welcomed the paper "Improved Probability Prediction Method Research for Photovoltaic Power Output" by Ze Cheng, Qi Liu, and Wen Zhang. In this interesting piece of work, the authors presented probabilistic prediction methods. The results showed that existing models can be improved to increase the predictability in photovoltaic plants power output.

Interestingly, most of the received papers touched, directly or indirectly, the use of alternative sources of energy for the transport sector. According to a recent International Energy Agency report [6], emissions from transport are continuing to rise, and several modes of transport appear complex to decarbonize. This topic is challenging, and many papers in this Special Issue present interesting analysis and technical solutions.

In "Is Deployment of Charging Station the Barrier to Electric Vehicle Fleet Development in EU Urban Areas? An Analytical Assessment Model for Large-Scale Municipality-Level EV Charging Infrastructures" by Giacomo Talluri, Francesco Grasso, and David Chiaramonti, the authors investigated the minimum charging infrastructure size and cost for two typical EU urban areas (Florence and Brussels). The analysis shows how policy can steer the deployment of infrastructures, especially with respect to the distribution of fast vs. slow/medium charging stations (CS). Interestingly, the authors pointed out that the critical barrier for CS development in the two urban areas is likely to become the time needed to install CS in the urban context, rather than the related additional electric power and costs.

Again on the possibility to use electricity to decarbonize transport modes, the Special Issue contains an interesting paper titled: "PTG-HEFA Hybrid Refinery as Example of a SynBioPTx Concept—Results of a Feasibility Analysis" by Franziska Müller-Langer, Katja Oehmichen, Sebastian Dietrich, Konstantin M. Zech, Matthias Reichmuth, and Werner Weindorf. The work analyses the aviation sector, considering that limited alternative fuels for a CO_2-neutral aviation sector have already been accepted by ASTM-certificarion process [7]. Among others, synthetic paraffinic kerosene from hydrotreated esters and fatty acids (HEFA–SPK) is a sustainable aviation fuel. This fuel can be produced via power-to-gas pathways, and alternative scenarios on feedstocks, electricity supply, necessary hydrogen supply, and different main products are analyzed. As a result, the attainment of at least 50% GHG mitigation might be possible: this highly depends on the renewability grade of the hydrogen provision as well as on the used feedstock. The scenario in which hydrogen is produced by steam reforming of internally produced naphtha proves to be the best combination of highly reduced GHG emissions and low HEFA–SPK production costs.

Energy storage can also be based on chemical molecules. Ethanol production from cellulosic material is considered one of the most promising options for future biofuel production contributing to both energy diversification and decarbonization of the transport sector [8], especially where electricity is not a viable option (e.g., aviation). In the paper "What is Still Limiting the Deployment of Cellulosic Ethanol? Analysis of the Current Status of the Sector" by Monica Padella, Adrian O'Connell, and Matteo Prussi, a comprehensive overview of the status of cellulosic ethanol production in the EU and outside the EU is presented. This was made by reviewing the available literature and highlighting technical and non-technical barriers that still limit cellulosic ethanol production at a commercial scale. The review shows that the cellulosic ethanol sector appears to be still stagnating, characterized by technical difficulties as well as high production costs. Competitiveness issues against standard starch-based ethanol are evident, considering many commercial-scale cellulosic ethanol plants appear to be currently in idle or on-hold states.

Beside road and aviation, maritime transport is a mode of transport that is seeking low-GHG alternatives [9]. In the paper "Analysis of Internal Gas Leaks in an MCFC System Package for an LNG-Fueled Ship" by Gilltae Roh, Youngseung Na, Jun-Young Park, and Hansung Kim, a 300 kW molten carbonate fuel cell (MCFC) for maritime application system was studied. The paper presented the challenge to ensure safety in case of a gas leak by applying computational fluid dynamics (CFD) techniques.

On the same subject, the paper "Analysis of a Supercapacitor/Battery Hybrid Power System for a Bulk Carrier" by Kyunghwa Kim, Juwan An, Kido Park, Gilltae Roh, and Kangwoo Chun, presented a hybrid power system combining conventional diesel generators with two different energy storage systems (ESSs) (lithium-ion batteries (LIB) and supercapacitors (SC)) focused on port operations of ships. The results show that the proposed system can reduce emissions (CO_2, SOx, and NOx) substantially and has a short payback period, particularly for ships that have a long cargo handling time or visit many ports with a short-term sailing time.

3. Perspectives on the Future of Research and Developments

All the interesting received contributions allowed covering a wide range of applications of alternative energy, expanding the original borders and definitions of the Special Issue. Although the Special Issue is now closed, more in-depth research in renewable energy technologies is expected. The papers received and the interaction with the author have encouraged us to propose a new Special Issue titled: "Frontier Trends of Renewable Energy Production and Storage Technologies".

This new Special Issue is looking for contributions on these topics, and MDPI and the editors of the journal *Applied Sciences* are delighted to have the privilege of publishing this Special Issue.

We would like to thank all the authors who contributed to the success of the Special Issue "Cutting-Edge Technologies for Renewable Energy Production and Storage" and we look forward to new interesting papers.

Acknowledgments: This issue would have not been possible without the contributions of various talented authors, hardworking and professional reviewers, and the dedicated editorial team of *Applied Sciences*. As the Editor, I would like to congratulate all authors. Finally, I would like to take this opportunity to record my sincere gratefulness to all reviewers and place on record the gratitude towards the editorial team of *Applied Sciences*, with special thanks to Damaris Zhao from MDPI Branch Office.

References

1. Carbajo, R.; Cabeza, L.F. Renewable energy research and technologies through responsible research and innovation looking glass: Reflexions, theoretical approaches and contemporary discourses. *Appl. Energy* **2018**, *211*, 792–808. [CrossRef]
2. Letcher, T.M. 11—Storing Electrical Energy. In *Managing Global Warming. An Interface of Technology and Human Issues*; Academic Press: New York, NY, USA, 2019; pp. 365–377.
3. Lippke, B.; Gustafson, R.; Venditti, R.; Volk, T.; Oneil, E.; Johnson, L.; Steele, P. Sustainable biofuel contributions to carbon mitigation and energy independence. *Forests* **2011**, *2*, 861–874. [CrossRef]
4. Müller-Langer, F.; Oehmichen, K.; Dietrich, S.; Zech, K.M.; Reichmuth, M.; Weindorf, W. PTG-HEFA Hybrid Refinery as Example of a SynBioPTx Concept—Results of a Feasibility Analysis. *Appl. Sci.* **2019**, *9*, 4047. [CrossRef]
5. Singh, G.K. Solar power generation by PV (photovoltaic) technology: A review. *Energy* **2013**, *53*, 1–13. [CrossRef]
6. IEA. Improving the Sustainability of Passenger and Freight Transport. 2019. Available online: www.iea.org/topics/transport (accessed on 16 January 2020).
7. EASA. *European Aviation Environmental Report 2019*; EASA, EEA, Eurocontrol: Brussels, Belgium, 2019.
8. Liu, C.-G.; Xiao, Y.; Xia, X.-X.; Zhao, X.-Q.; Peng, L.; Srinophakun, P.; Bai, F.-W. Cellulosic ethanol production: Progress, challenges and strategies for solutions. *Biotechnol. Adv.* **2019**, *37*, 491–504. [CrossRef] [PubMed]
9. The 70th Session of the Marine Environment Protection Committee. Amendments to the Annex of the Protocol of 1997 to Amend the International Convention of Pollution from Ships, 1973, as Modified by the Protocol of 1978 Relating Thereto. In *Proceedings Resolution MEPC.278 (70)*; IMO: London, UK, 2016.

Article

Improved Probability Prediction Method Research for Photovoltaic Power Output

Ze Cheng *, Qi Liu * and Wen Zhang *

School of Electrical Engineering and Automation, Tianjin University, Tianjin 300000, China
* Correspondence: chengze@tju.edu.cn (Z.C.); liuqi667@foxmail.com (Q.L.); wenzhang@tju.edu.cn (W.Z.)

Received: 5 April 2019; Accepted: 8 May 2019; Published: 17 May 2019

Abstract: Due to solar radiation and other meteorological factors, photovoltaic (PV) output is intermittent and random. Accurate and reliable photovoltaic power prediction can improve the stability and safety of grid operation. Compared to solar power point prediction, probabilistic prediction methods can provide more information about potential uncertainty. Therefore, this paper first proposes two kinds of photovoltaic output probability prediction models, which are improved sparse Gaussian process regression model (IMSPGP), and improved least squares support vector machine error prediction model (IMLSSVM). In order to make full use of the advantages of the different models, this paper proposes a combined forecasting method with divided-interval and variable weights, which divides one day into four intervals. The models are combined by the optimal combination method in each interval. The simulation results show that IMSPGP and IMLSSVM have better prediction accuracy than the original models, and the combination model obtained by the combination method proposed in this paper further improves the prediction performance.

Keywords: PV; probability prediction; sparse Gaussian process regression; least squares support vector machine; combination method

---

1. Introduction

At present, the average natural gas storage in the world is 53 years. There is more coal in storage than oil and natural gas, and the world's coal storage capacity is 15,980 tons, which can be mined for about 200 years [1]. It can be seen that a shortage of fossil fuels is imminent. Photovoltaic (PV) power generation has developed rapidly in the world in recent years due to its advantages in meeting energy demands, reducing environmental pollution, and improving energy structure [2]. As a result of solar radiation and other factors, PV power output has high volatility and randomness, and with the increase of photovoltaic grid-connected capacity, this adverse impact will bring more and more risks to grid operation. Therefore, accurate prediction of photovoltaic power generation will be of great significance to the stability and safety of grid dispatching and power systems [3].

In many previous studies, the prediction of photovoltaic power generation based on physical models has shown great progress. Zhang et al. [4] established the basic model of the photovoltaic cell and photoelectric conversion efficiency based on the principle of photovoltaic power generation and photoelectric conversion efficiency model. Then empirical formulas affecting photoelectric conversion efficiency were obtained, and reasonable empirical parameters were selected to predict PV output power. Dolara et al. [5] proposed the three-parameter, four-parameter, and five-parameter equivalent circuit model comparison method for photovoltaic cells and two thermal models for estimating battery temperature, which can achieve better accuracy with fewer parameter solutions. However, the creation of physical model and parameter solving process are complicated, and the anti-interference ability of the model is poor.

With the development of machine learning technology and computer hardware capabilities, an increasing number of machine learning and statistical regression methods have been applied to the field of photovoltaic power generation prediction. This includes artificial neural network algorithms [6], support vector machine (SVM) algorithms [7,8], the ARMAX algorithm [9], the Markov chain algorithm [10], and so on. Chen et al. [11] proposed a neural network-based photovoltaic power generation prediction model that can predict power generation under different weather conditions one day in advance. Chen et al. [12] proposed a solar irradiance prediction method based on the support vector machine algorithm. This method uses different kernel functions to predict solar irradiance, which are then compared. Based on the ARMAX algorithm, Li et al. [13] considered meteorological factors to obtain more-accurate prediction results. A photovoltaic power generation prediction model based on an improved Markov chain was proposed by Ding [14]. The Markov chain is mainly used to correct the residual of the prediction model to improve accuracy. However, these algorithms have obvious defects. It is easy for the neural network algorithm to fall into the local minimum, and the model is not well explained. The support vector machine has limited processing ability for large samples of data, and the time series method has weak non-stationary processing ability. Many scholars have improved the efficiency of models by improving the algorithm. Eseye et al. [15] used the particle swarm optimization (PSO) to optimize the normalization parameters and kernel parameters of SVM. The back propagation (BP) neural network algorithm was improved by combining the momentum term with the variable learning rate, so the defects—the traditional BP learning algorithm is liable to fall into local minimum points, and has a slow convergence rate—were remedied. The improved BP neural network is used to improve the predictive performance [16].

The output of PV generation is restricted by its external environment, and the weather has a greater impact on photovoltaic systems. In order to reduce the impact of weather, data are classified according to the type of the weather. Some results prove that such methods have a better performance [17,18]. In most models, only conventional variables such as temperature, humidity, and wind speed are generally considered, so the accuracy of these models will decrease under extreme weather conditions. The aerosol index (AI) can indicate that there is a strong linear relationship between particulate matter in the atmosphere and solar radiation attenuation, which has a potential impact on the energy generated by photovoltaic panels. In [19], based on the classification of seasons, AI can be used as an additional input parameter to adapt to the complicated environment. The drawback of these methods, then, is that the accuracy of models depends largely on the accuracy of the weather classification.

The accuracy of predictive models based on statistical learning depends mainly on a large amount of historical data. However, historical data contains complex information, and there is redundant information that may not be necessary. Not all weather factors have a significant impact on PV output, so we need to extract or remove information from historical data to reduce the complexity of models. In [20], it is shown that temperature and insolation are positively correlated with PV power, humidity is negatively correlated with PV power, and wind speed has no obvious correlation with PV power by the correlation analysis. Therefore, air temperature, humidity, insolation, and historical PV power can be selected as inputs to the prediction model to reduce the complexity of the data. Zhu et al. [21] proposed a PV output prediction method combining wavelet decomposition and the artificial neural network (ANN) algorithm. After separating the useful information and the interference information by wavelet decomposition, the neural network model is used to obtain the predicted power value. Malvoni et al. [22] combined the quadratic Renyi entropy criteria with principal component analysis (PCA) to reasonably reduce the data dimension and use least squares support vector machines (LSSVM) to predict future PV power. This model can facilitate calculation, while improving accuracy. In addition, it has been found that the introduction of image data can also improve prediction accuracy. Marquez et al. [23] proposed a method for combining solar cloud image data and artificial neural network models to predict solar irradiance. Zhu Xiang et al. [24] used cloud information and cloud maps in numerical weather prediction (NWP) to predict the power attenuation caused by cloud clusters blocking photovoltaic power plants over the subsequent 4 hours, then corrected the predicted values to improve the accuracy of model prediction. This type of method requires more advanced experimental equipment, however.

Most of the research has aimed to predict a certain value at a certain moment, but it is difficult for point prediction values to express the uncertainty of the prediction result, which will affect power grid scheduling and the stability of the power system. Compared with point prediction, probabilistic prediction makes up for the shortcoming that point prediction cannot measure the uncertainty of prediction results [25]. Due to the uncertainty of solar resources and the inherent defects of the prediction model, the point prediction error of solar power cannot be avoided, and the defect that the point prediction result cannot make a quantitative description of the uncertainty of solar power is difficult to overcome. In terms of the application of solar power, there needs to be a relatively accurate estimation of the fluctuation range of solar power, which requires planning, operation, safety, and stability analysis of the power grid (including solar power generation). The probabilistic prediction of solar power generation expands the connotations of solar power generation prediction, and can provide the probability distribution of PV power generation. The diversity of probability distribution at different time points can provide power system policymakers with an abundance of uncertain information, including economic dispatch, rotating standby arrangement, and electricity market price optimization problems. In a word, the probabilistic prediction method can give the possible PV power value and its probability distribution in the next moment, and provide more-comprehensive prediction information [26]. However, the current research on probability prediction in the field of photovoltaic power generation is still in its infancy. Fatemi et al. [27] proposed two parametric probability prediction methods for predicting solar irradiance by β-distribution and bilateral power distribution, effectively predicting solar irradiance and accurately describing its stochastic characteristics. Fonseca et al. [28] assumed the prediction error distribution as the normal distribution and the Laplacian distribution. The probability distribution of the generated power and the confidence interval value at different confidence levels was then obtained by the maximum likelihood estimation method. Almeida et al. [29] used the meteorological data obtained by NWP as input data, and a probability prediction model based on a quantile regression prediction algorithm was established to study the probability prediction of photovoltaic power generation. Mohammad et al. [30] combined the probability distribution theory with the Gaussian mixture method, and the prediction results are consistent with the actual probability distribution of photovoltaic power under different weather conditions.

Considering that PV output is affected by many weather factors, and that the complexity of a model will increase if too many variables exist, this paper preprocesses the original data set based on feature selection and similar sample classifications. It is difficult for the point prediction method to express the uncertainty of prediction results. Two improved photovoltaic power generation probabilistic prediction models are proposed in this paper to improve the preliminary prediction performance. The idea of traditional combination methods is to combine different algorithms to optimize a certain prediction model without changing the essential structure of the models. Through analysis, it was found that different models had different advantages during different periods of time. In order to improve the prediction accuracy by contributing to the advantages of the different models, a new probability prediction model with multi-interval and variable weight is proposed. The method is applied to handle the uncertainties of photovoltaic power generation for the first time, to the best of our knowledge.

The major contributions and innovations of this paper are as follows:

1. The improved grey wolf optimization algorithm is used to optimize the sparse Gaussian process model and the least squares support vector machine model. The better super-parameter values are obtained, which improve accuracy.

2. An error correction probability prediction model based on improved least squares support vector machine (IMLSSVM) is proposed. The model can be corrected by using the historical error distribution to realize the probability prediction model under the premise of realizing point prediction.

3. A piecewise optimal combination model is proposed that uses different combined weighting coefficients in different periods to make full use of different prediction models.

2. Single Improved Prediction Model

In this part, two improved prediction models will be introduced. The Gaussian process (GP) is a generalization of the Gaussian probability distribution. The sparse Gaussian process regression (SPGP) model is a supervised learning model based on Bayesian theory and statistical theory. Least squares support vector machine (LSSVM) is a machine learning algorithm based on statistical learning theory. In order to obtain the optimal solution of the super-parameters, the improved grey wolf optimization algorithm was used to solve the super-parameters of the SPGP model and LSSVM. In this paper, improved SPGP model and LSSVM model are used to predict photovoltaic power, respectively.

2.1. Improved Sparse Gaussian Process Regression Model

2.1.1. Sparse Gaussian Process Regression

In the Gaussian process (GP) [31] model, it is assumed that the prior distribution of the observed values is $\mathbf{y}$ given by a Gaussian whose mean is zero and whose covariance is defined by a Gram matrix $\mathbf{K} + \sigma_n^2 \mathbf{I}$ so that

$$\mathbf{y} \sim N(0, \mathbf{K} + \sigma_n^2 \mathbf{I}) \tag{1}$$

where $\mathbf{K} = k(\mathbf{x}_n, \mathbf{x}_n)$ with k as the kernel function. For a new input $\mathbf{X}_*$, a prediction of the target variable $\mathbf{f}_*$ can be made by using the rules for conditioning Gaussians. The predictive distribution $\mathbf{f}_*$ is a Gaussian distribution with mean $\overline{\mathbf{f}_*}$ and covariance $\mathrm{cov}(\mathbf{f}_*)$. The posterior distribution of the predicted values is

$$\mathbf{f}_* | \mathbf{X}, \mathbf{y}, \mathbf{X}_* \sim N(\overline{\mathbf{f}_*}, \mathrm{cov}(\mathbf{f}_*)) \tag{2}$$

where $\overline{\mathbf{f}_*} = E[\mathbf{f}_* | \mathbf{X}, \mathbf{y}, \mathbf{X}] = \mathbf{K}(\mathbf{X}_*, \mathbf{X})[\mathbf{K}(\mathbf{X}, \mathbf{X}) + \sigma_n^2 \mathbf{I}]^{-1} \mathbf{y}, \mathrm{cov}(\mathbf{f}_*) = \sigma_n^2 + \mathbf{K}(\mathbf{X}_*, \mathbf{X}_*) - \mathbf{K}(\mathbf{X}_*, \mathbf{X})(\mathbf{K}(\mathbf{X}, \mathbf{X}) + \sigma_n^2 \mathbf{I})^{-1} * \mathbf{K}(\mathbf{X}, \mathbf{X}_*)$.

In this paper, the commonly used automatic relevance determination function is chosen as the covariance function.

$$k(x_p, x_q) = \sigma_f^2 \exp\left(-\frac{1}{2l^2}\|x_p - x_q\|^2\right) + \sigma_n^2 \delta_{pq} \tag{3}$$

where x_p, x_q are the variables of training or test sets, l, σ_f, σ_n are the hyper-parameters, and δ_{pq} is the symbolic function.

A main drawback of GP is its high time complexity. A training process requires inverting a matrix $\mathbf{K} + \sigma_n{}^2 \mathbf{I}$, which costs $O(n^3)$. In order to solve this problem, in this paper, an SPGP model that adopts the fully independent training conditional approximation (FITC) approximation method [32] is introduced. The FITC method adopts a pseudo dataset D and uses M samples to approximately simulate the original dataset, where the number of samples M $<<$ N. So the posterior distribution $\overline{\mathbf{f}}$ over pseudo targets $\overline{\mathbf{X}}$ can be obtained.

$$p(\overline{\mathbf{f}}|D, \overline{\mathbf{X}}) = N(\mathbf{K}_M \mathbf{Q}_M^{-1} \mathbf{K}_{MN} (\Lambda + \sigma^2 \mathbf{I})^{-1} \mathbf{y}, \mathbf{K}_M \mathbf{Q}_M^{-1} \mathbf{K}_M) \tag{4}$$

where $[\mathbf{K}_M]_{mm'} = K(\overline{x}_m, \overline{x}_{m'})$, $\mathbf{Q}_M = \mathbf{K}_M + \mathbf{K}_{MN}(\Lambda + \sigma^2 \mathbf{I})^{-1} \mathbf{K}_{NM}$, $[\mathbf{K}_{MN}]_{mn} = K(x_m, \overline{x}_n)$, and $\Lambda = diag(\lambda)$, $\lambda_n = K_{nn} - k_n^\top K_M^{-1} k_n$. Finally, predictive distribution of test data y_* can be obtained and it is a Gaussian distribution with mean μ_* and covariance σ_*^2.

$$p(y_* | \mathbf{x}_*, D, \overline{\mathbf{X}}) = \int p(\overline{\mathbf{f}}|D, \overline{\mathbf{X}}) p(y_* | \mathbf{x}_*, \overline{\mathbf{X}}, \overline{\mathbf{f}}) d\overline{\mathbf{f}} = N(\mu_*, \sigma_*^2) \tag{5}$$

where $\mu_* = \mathbf{k}_*^\top \mathbf{Q}_M^{-1} \mathbf{K}_{MN}(\Lambda + \sigma^2 \mathbf{I})^{-1} \mathbf{y}, \sigma_*^2 = \mathbf{K}_{**} - \mathbf{k}_x^\top (\mathbf{K}_M^{-1} - \mathbf{Q}_M^{-1}) \mathbf{k}_* + \sigma^2, \sigma_*^2 = \mathbf{K}_{**} - \mathbf{k}_x^\top (\mathbf{K}_M^{-1} - \mathbf{Q}_M^{-1}) \mathbf{k}_* + \sigma^2$.

The matrix $\mathbf{K}_{MN}(\Lambda + \sigma^2 \mathbf{I})^{-1} \mathbf{K}_{NM}$ is $O(M^2 N)$. Since M is much less than N, the computational costs are lower than those of the GP.

2.1.2. Improved Grey Wolf Optimization Algorithm

Because the SPGP model is complicated and the concavity and convexity of the optimization problem formula (Equation (7) cannot be judged intuitively, traditional convex optimization methods may not be applicable to the solution of the hyper-parameter in the SPGP model. Moreover, because this method relies too heavily on initial values, it falls easily into local optimum. Therefore, this paper introduces an improved grey wolf optimization algorithm to optimize the hyper-parameter. Grey wolf optimization (GWO) is a heuristic population intelligent optimization algorithm proposed by Mirjalili in 2014 [33]. It has the advantages of simple principles, less parameters and a strong global search ability, but there is still much room for improvement in the algorithm.

When dealing with the optimization problem, it is assumed that in the D-dimensional search space, the number of grey wolves is N, and the position of the i-th wolf is defined as $X_i = (X_i^1, \cdots, X_i^D)$, where X_i^k indicates the position of the i-th wolf in the k-th dimension. This paper will improve the algorithm to obtain an improved grey wolf optimization (IMGWO) according to the following three aspects:

(1) Opposition-based learning algorithm: The initial population is obtained by randomly generated methods in the GWO algorithm. However, the initial population will affect the search efficiency and the quality of the solution. Therefore, this paper adopts the method of opposition-based learning to generate a diverse initial population. Assuming that there is a number x in $[m, n]$, the opposite point of x is defined as $x' = m+n-x$.

(2) Nonlinear convergence factor: A common problem facing optimization algorithms is how to balance a global search ability with a local search ability. In GWO, the convergence factor decreases linearly from 2 to 0 over the course of iterations. It cannot reflect the actual search process because the process is not linear. Therefore, in this paper, a new non-linear convergence factor is proposed: $\vec{a} = 2 - 2(k/K))^n$, where k indicates the current iteration, K is the maximum number of iterations, n is the attenuation order, and n is 3.

(3) Dynamic weight updating: In the stage of position updating, GWO uses an equal weight method that ignores the different characteristics of alpha, beta, and delta. The wolves have different importance for the population, so we need different weights to reflect the different importance. In this paper, fitness values are used to describe the importance of wolves. Equation (6) is adopted to update location information:

$$\vec{X}(t+1) = w_\alpha \vec{X}_1 + w_\beta \vec{X}_2 + w_\sigma \vec{X}_3 \tag{6}$$

where $w_i = f(X_i(t))/(f(X_\alpha(t))+f(X_\beta(t))+f(X_\delta(t)))(i = \alpha, \beta, \delta)$ is the weight coefficient, $f(X_i(t))$ is the fitness value, and $\vec{X}_i(i = \alpha, \beta, \delta)$ is the position of the wolf.

In this paper, the flow chart of the GWO algorithm is given, as shown in Figure 1.

The main difficulty of SPGP lies in the solution of the hyper-parameters. In order to obtain the hyper-parameters, this paper adopts the sum of root mean square error (RMSE) and continuous ranking probability score (CRPS) as a target fitness. The fitness is shown as:

$$\begin{cases} \min(RMSE + CRPS) \\ s.t.\, 0 \leq RMSE \leq 1 \\ \quad 0 \leq CRPS \leq 1 \end{cases} \tag{7}$$

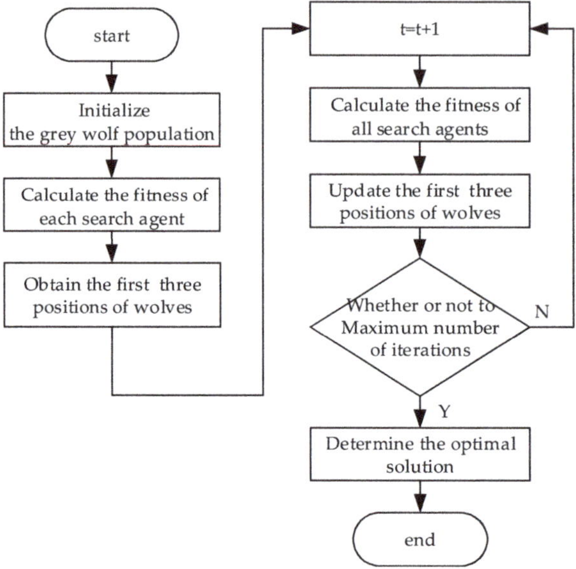

Figure 1. The flow chart of grey wolf optimization algorithm.

2.2. Improved Least Squares Support Vector Machine Error Prediction Model

2.2.1. Improved Least Squares Support Vector Machine

Because of the quadratic programming problem under the constraint of inequality in the support vector machine, the calculation amount is large. In order to improve the calculation speed, Vandewallw et al. [34] proposed LSSVM in 1999, and converted the inequality constraint into an equality constraint. The complex convex quadratic programming problem is transformed into a linear equation solution. Its optimization objective function is as follows:

$$\min\{\tfrac{1}{2}\mathbf{w}\cdot\mathbf{w} + \tfrac{C}{2}\sum_{i=1}^{n}\varepsilon_i^2\}$$
$$s.t. y_i = \mathbf{w}\cdot\varphi(\mathbf{x}_i) + b + \varepsilon_i, i = 1,2\cdots,n \tag{8}$$

where $\mathbf{w}$ is the weight coefficient vector, C and b are constants, and ε_i is the error.

In order to solve the minimization problem, the Lagrangian function method is used.

$$L(\mathbf{w},b,\varepsilon_i,\alpha) = \frac{1}{2}\mathbf{w}\cdot\mathbf{w} + \frac{C}{2}\sum_{i=1}^{n}\varepsilon_i^2 - \sum_{i=1}^{n}\alpha_i[\mathbf{w}\cdot\varphi(\mathbf{x}_i)+b+\varepsilon_i-y_i] \tag{9}$$

where α is Lagrangian factor.

Then we need to take the derivative of these parameters:

$$\frac{\partial L}{\partial \mathbf{w}} = \mathbf{w} - \sum_{i=1}^{n}\alpha_i\varphi(x_i) = 0$$
$$\frac{\partial L}{\partial b} = \sum_{i=1}^{n}\alpha_i = 0$$
$$\frac{\partial L}{\partial \varepsilon_i} = C\varepsilon_i - \alpha_i = 0 \tag{10}$$
$$\frac{\partial L}{\partial \alpha_i} = \mathbf{w}\cdot\varphi(\mathbf{x}_i) + b + \varepsilon_i - y_i = 0$$

The LSSVM model can be obtained by solving the above equation, which can be expressed as

$$f(x) = \sum_{i=1}^{n} \alpha_i K(x_i, x) + b \tag{11}$$

where $K(x_i, x)$ is the kernel function.

This paper implements the optimization process of the LSSVM using the IMGWO algorithm.

2.2.2. Error Probability Distribution of Power Prediction

If the potential law of the prediction error can be properly expressed, then the prediction value can be corrected to improve the prediction performance. In this paper, the prediction values are obtained via LSSVM, and the corresponding prediction error values are be obtained by calculation. The methods for estimating the distribution of error can be divided into parameter estimation and non-parametric estimation. The parameter estimation assumes that the data sample obeys a known distribution function, and that the samples are used to estimate the parameter values of the distribution function. Normal distribution, T-location-scale distribution, and logistic distribution are the distribution functions commonly used. Non-parametric estimation is based on the characteristics of the sample data, not a priori estimation of data samples. The entire estimation process is completely driven by the data samples, and has a strong utility to the problem which is impossible to estimate the characteristics of its sample data in advance. Kernel density estimation (KDE) [35] is chosen as the nonparametric estimation method in this paper. $X_1, X_2, \cdots, X_n$ is the overall sample, and $x_1, x_2, \cdots, x_n$ represents the observed values of samples. The kernel density estimate of the probability density function at any point x is defined as

$$f_h(x) = \frac{1}{nh} \sum_{i=1}^{n} K(\frac{x - X_i}{h}) \tag{12}$$

where $f_h(x)$ is the estimated probability density function, h is the window width, and $K(\cdot)$ is the kernel function.

Through statistical analysis of the error between the actual and predicted values of photovoltaic power generation, we find that the prediction errors in different power intervals have different distributions, which is quite different from the distribution of the overall prediction error. Therefore, this paper divides the data into multiple intervals, and performs statistical analysis on the error of each interval to calculate the probability density function. When the sample data is relatively sufficient, the empirical distribution is similar to the overall distribution, and it can be treated as the total distribution.

For the division of intervals, there are no fixed rules to follow. If the number of the intervals is too large, the amount of sample data in each interval will be small. The empirical distribution will be different from the overall distribution, which cannot reflect the real information of sample data. On the contrary, the intrinsic information and the meaning of segmentation statistics will be lost. Assuming that the length of the interval is ΔP and the power interval range is $[P_1, P]$, the power is divided at equal intervals.

$$D_i = [P_1 + (i-1)\Delta P, P_1 + i\Delta P] \tag{13}$$

where $i = 1, \cdots, N$, and N is the number of intervals.

By dividing the range of each interval by Equation (13), there may be cases where the number of samples in some intervals is too small. In order to better reflect the distribution of data, it is necessary to adjust the range of the interval to a certain extent, according to the actual situation.

3. Probability Prediction Combination Model

The essence of the combination method is to find the respective superior information in a variety of models that can more fully describe the sample information than single model. Therefore, it, with a stronger anti-interference ability, can effectively reduce the influence of complex environmental factors [36]. It should be noted that the performance of any single model directly effects the accuracy of the combined model. Considering the calculation cost and effect comprehensively, the number of single models is often taken as two to five. In addition, the quantile regression neural network (QRNN) model has better performance in predicting PV output [37]. This model and the proposed two improved models will be used to construct the combined model in this paper. In this part, a piecewise optimal combination model is proposed that uses different combined weighting coefficients in different periods to make full use of different prediction models.

3.1. Optimal Combination Method

The optimal combination method is based on the functional relationship between the combination model and the single model. The objective function is constructed under certain constraints, and the weight coefficient of the single models in the combination method is solved by maximizing the objective function.

The non-optimal combination method mainly follows the simple and convenient principle of obtaining the weight coefficient using methods such as the arithmetic average method, the root mean square error reciprocal method, the entropy method, and so on. The accuracy of the non-optimal combination method is generally lower than that of the optimal combination method. In this paper, the interpretation of the optimal linear combination method is given by taking the sum of squared prediction errors as the objective function to solve the weight coefficient.

If x_t is the true value of the t-th sample, x_{it} is the predicted value of the t-th sample in the i-th single model, and e_{it} is the error value of the t-th sample of in the i-th prediction model, then $e_{it} = x_t - x_{it}$. In order to ensure the validity and unbiasedness of the results, the weight coefficient needs to meet the following conditions:

$$\sum_{i=1}^{m} w_i = 1 \tag{14}$$

where w_i represents the weight coefficient of the i-th single model. The sum of the squared error of the combination model can be expressed as follows:

$$J = \sum_{t=1}^{n} e_t^2 = \sum_{t=1}^{n} \sum_{i=1}^{m} \sum_{j=1}^{m} w_i e_{it} w_j e_{jt} \tag{15}$$

Therefore, the linear combination model can use the sum of the squared error as the objective function. If $W = (w_1, \cdots, w_m)^T$, $E_{ij} = e_i^T e_j = \sum_{i=1}^{N} e_{it} e_{jt}$, $E = (E_{ij})_{n \times n}$, R is an n-dimensional column vector where each element is 1, and E is a square matrix of $n \times n$, which represents the information matrix of the combined prediction error, then the solution of the weight coefficient can be expressed as an optimization problem.

$$\begin{aligned} \min J &= \sum_{t=1}^{n} e_t^2 = L^T E L \\ s.t. R^T L &= 1 \\ L &\geq 0 \end{aligned} \tag{16}$$

The solution is described in detail in [38]. For the solution of the weight coefficient in the optimal combination, this paper uses the IMGWO proposed in this paper to solve the problem. The position of wolves is the value of the weight coefficient.

3.2. Combination Method with Variable Intervals and Weights

In order to fully utilize the characteristics of each model at different time periods, a combination method with different weights of intervals is proposed. For example, the accuracy of method A at interval 1 is higher than that of method B, and the prediction accuracy at interval 2 is lower than that of method B, so method A and method B will have different weight coefficients in different time intervals.

When the day is divided into H intervals, the corresponding combination model will be respectively established. In this paper, the daily time period is divided into four intervals. Interval I is from 06:00 to 09:00, interval II is from 09:00 to 12:00, interval III is from 12:00 to 15:00, and interval IV is from 15:00 to 18:00. In the actual experiment, we first need to determine which time interval the predicted data belongs to, then use the combination model to predict the power generation. In this paper, a flow chart of the combination method is given, shown in Figure 2.

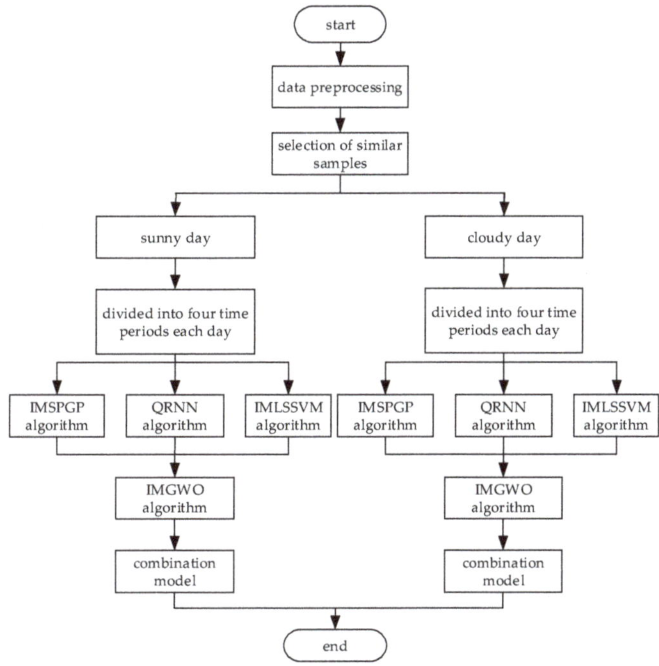

Figure 2. The flow chart of combination algorithm.

4. Experiment

4.1. Data Preprocessing

The dataset used in this paper is acquired from solar power station, from 2012 to 2014, of the Energy Forecasting Working Group of the Institute of Electrical and Electronics Engineers, including the 24-hour PV output value of photovoltaic panels and corresponding meteorological parameters such as surface solar irradiance, atmosphere, irradiance, surface pressure, total cloud amount, and so on [39]. Since the output of the photovoltaic panel was 0 at night, the data from 6:00 to 18:00 every day is selected as the raw data.

$$\overline{x}_{ij} = \frac{x_{ij} - x_{\min,j}}{x_{\max,j} - x_{\min,j}} \tag{17}$$

where x_{ij} is the i-th data of the j-th variable in the original dataset; $x_{\min,j}$ and $x_{\max,j}$ are the minimum and maximum values of the j-th variable in the original dataset, respectively; and $\overline{x}_{ij}$ is the i-th new data of the j-th variable after the normalization. The target variable values were normalized by the Energy Forecasting Working Group, so its rang is [0,1]. The other 12 variables provided were not normalized, and had different dimensions. After the normalization, the range of 12 variables are all [0,1].

If all variables are considered, it will inevitably increase the model complexity and training time. The random forest algorithm can solve the high-dimensional data classification and regression problem [40,41]. This paper uses the algorithm to select the variables affecting the PV output value. The variable description and processing results are shown in Table 1.

Table 1. Variable description.

Name	Unit	Score
Total column liquid water (TCLW)	kg·m^{-2}	13.2
Total column ice water (TCIW)	kg·m^{-2}	19.4
Surface pressure (SP)	Pa	15.1
Relative humidity at 1000 mbar (RH)	%	62.6
Total cloud cover (TCC)	0–1	16.8
10 meter U wind component (10 U)	m·s^{-1}	21.7
10 meter V wind component (10 V)	m·s^{-1}	13.3
2 meter temperature (2T)	K	39.7
Surface solar rad down (SSRD)	J·m^{-2}	370.3
Surface thermal rad down (STRD)	J·m^{-2}	50.7
Top net solar rad (TSR)	J·m^{-2}	208.0
Total precipitation (TP)	M	11.4

The 12 variables obtained from the European Centre for Medium-Range Weather Forecasts (ECMWF) NWP output are independent. Total column liquid water (TCLW) represents the vertical integration of cloudy liquid water content. Relative humidity at 1000 mbar (RH) is defined with respect to the saturation of the mixed phase (i.e., with respect to saturation over ice below −23 °C, and with respect to saturation over water above 0 °C). Total cloud cover (TCC) is derived from model levels using the model's overlap assumption. Top net solar rad (TSR) is the net solar radiation at the top of the atmosphere. Surface solar rad down (SSRD), surface thermal rad down (STRD), top net solar rad (TSR), and total precipitation (TP) are all accumulated data.

The importance scores of 12 meteorological parameters can be obtained by the random forest algorithm. The importance scores of the meteorological factors are shown from Table 1, and the importance ranking is performed according to the scores of the respective variables. A variable is more important when its score is higher. If there are too few variable factors left in the feature selection process, important information may be lost and the prediction accuracy may be affected. Therefore, the first six feature variables are selected as the input variables set through many tests. The six variables are SSRD, TSR, RH, STRD, 2T, 10U.

The photovoltaic power generation will show different trends under different weather conditions, and the prediction accuracy will be improved by classifying the original data and selecting similar sample. A fuzzy clustering algorithm can deal with similarity and fuzzy relations, so the algorithm is used to further process the data. On the one hand, too many classification types will make the number of samples for each type too small. On the other hand, PV output usually shows instability in the case of bad weather such as rainy days. The data are divided into sunny and rainy days by a fuzzy C-means clustering algorithm. The algorithm theory and calculation process are described in [20].

4.2. Evaluation Criteria

This paper introduces the evaluation criteria of point prediction value and confidence interval prediction value, which can evaluate the prediction probability distribution performance.

1) The root mean square error is defined as follows:

$$RMSE = \sqrt{\frac{\sum\limits_{i=1}^{N} (y_i - y_i')^2}{N}} \tag{18}$$

where N indicates the number of test set data, and y_i and y'_i represent the i-th predicted value and the actual value, respectively. The smaller the root mean square error value is, the higher the prediction accuracy of the model.

2)　The continuous ranking probability score is a comprehensive criterion for evaluating probability prediction performance. It can simultaneously evaluate the calibration degree and sharpness of the probability distribution. The smaller the value is, the better the prediction effect of the model will get. CRPS is defined as follows:

$$CRPS(F,x) = \begin{cases} \int_{-\infty}^{\infty} \{F(y)-1\}^2 dy, \ y \geq x \\ \int_{-\infty}^{\infty} (F(y))^2 dy, \ y < x \end{cases} \tag{19}$$

where x is the actual observed value and F is the predicted cumulative distribution function.

3)　The confidence interval average width (PINAW) represents the average of the confidence interval widths. The smaller the width value gets, the higher the accuracy. PINAW is defined as follows:

$$PINAW = \frac{\sum_{i=1}^{N} (up - down)}{N} \tag{20}$$

where up is the upper bound of the confidence interval, $down$ is the lower bound of the confidence interval, and N is the number of data in the test set.

4)　The prediction interval coverage (PICP) is the ratio of the number of target values falling within the confidence interval to the total number of predicted samples. The value ranges from 0 to 1. The closer the value is to 1, the better the prediction performance of the model is. PICP is defined as follows:

$$PICP = \frac{\gamma^{1-\alpha}}{N} \times 100\% \tag{21}$$

where $\gamma^{1-\alpha}$ is the number of target values that fall within the confidence interval.

When the PINAW becomes smaller, the PICP becomes larger, and the prediction performance is better. CRPS is the evaluation criterion that includes coverage and interval width, which can comprehensively reflect the performance of models.

4.3. Experiment and Analysis

4.3.1. Analysis of Experiment Results for the Single Models

Increasing the similarity between the training sample and the test sample can indirectly improve the prediction accuracy of the model. The weather conditions in the same months every year were similar, so the data set was divided into four quarters. In the experiment, one-month data for each quarter was selected as the test data set, and the three-month data in the same quarter of the previous year was used as the training data set. For example, the data from 1 to 31 May 2014 were used as test data, and the data from April, May, and June 2013 as training data.

Since the prior assumption of the GP model conforms to the normal distribution, the normal distribution theory can be used to calculate the probability prediction result of the photovoltaic power. The mean and variance of the variable to be predicted can be obtained by the GP model. The mean is used to indicate the prediction. The mean is used to represent the predicted value of the variable. Combined with the mean and the variance, we can obtain the probability prediction results, so as to obtain the upper and lower limits of the confidence interval at different confidence levels. The confidence interval can be expressed as follows:

$$\left(\mu - \frac{\sigma}{n}z_{\alpha/2}, \mu + \frac{\sigma}{n}z_{\alpha/2}\right) \tag{22}$$

where μ is the mean value, σ is the standard deviation, and $z_{\alpha/2}$ can be obtained by looking up the table. If the confidence level is 0.95, $z_{\alpha/2} = 1.96$.

In this paper, the IMGWO is compared with the GWO and PSO. The results are shown in Figure 3. The fitness value and iteration number of the IMGWO are 0.1065 and 25 times, respectively. The fitness value and iteration number of the GWO are 0.1086 and 38 times, respectively. The fitness value and iteration number of the PSO are 0.1146 and 18 times, respectively. Therefore, the IMGWO has a faster convergence speed and better fitness than the GWO. Although the convergence speed of the PSO is better than the IMGWO, the fitness value is larger than that of the IMGWO, indicating that the PSO falls into the local optimum. From the above analysis, it can be proven that the IMGWO has a stronger global search ability and better search speed. Figure 4 shows the IMSPGP model prediction results of seven days selected from May. The first four days are sunny, the last three days are cloudy, and the shaded parts represent 95% confidence intervals, 80% confidence intervals, and 50% confidence intervals, respectively. The photovoltaic output shows with certain regularity that the point prediction result is more accurate, the prediction interval is slightly wider at the peak, and the overall prediction accuracy is higher.

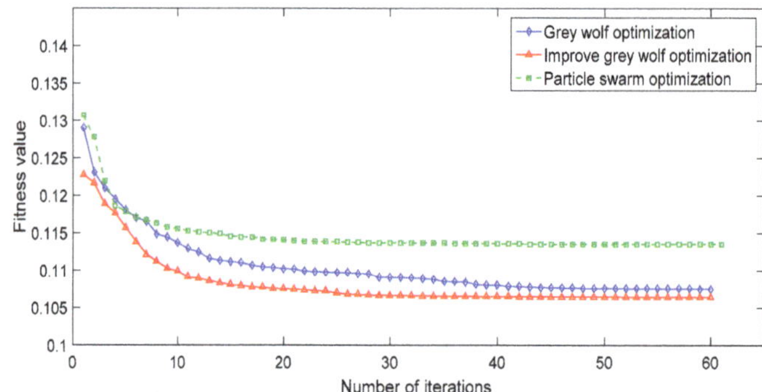

Figure 3. Comparison of fitness values.

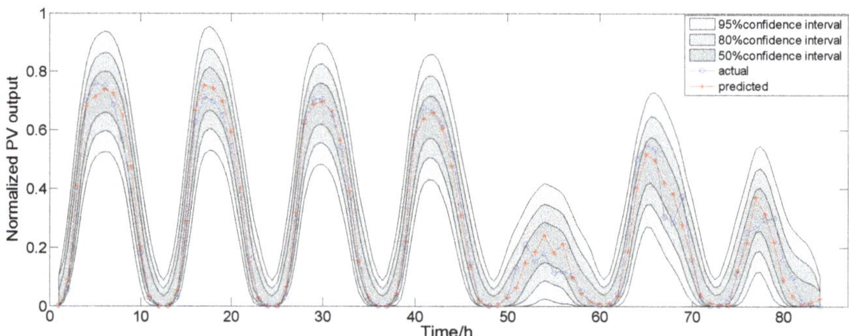

Figure 4. Point forecast and interval forecast over seven days using IMSPGP.

As can be seen from Table 2, IMSPGP had the smallest RMSE and CRPS, with optimal prediction performance, whether representing a sunny or rainy day. It should be explained that although the PISAW of the IMSPGP model is larger than that of the GP model and the SPGP model for sunny days, the PICP value is larger than that of the other two models. In this case, it is necessary to consider the size of the comprehensive evaluation criterion CRPS. Because the IMPSGP model has the smallest CRPS, it is shown that the model does improve the prediction effect.

To obtain the probability prediction results of IMLSSVM, this paper first calculates the prediction error of the IMLSSVM model for the whole year of 2012. The model is tested in different seasons. As can be seen from Table 3, the IMLSSVM has better predictive performance than the LSSVM. In terms of point prediction performance among the models, IMSPGP generally has the highest accuracy. As a whole, the IMLSSVM model is better than the QRNN model. In order to increase the diversity of the model, the data of the error distribution are obtained by using the IMLSSVM instead of the IMSPGP.

Table 2. Prediction results of GP, SPGP, and IMSPGP models for May.

Model	Sunny				Rainy			
	RMSE	**PINAW**	**PICP**	**CRPS**	**RMSE**	**PINAW**	**PICP**	**CRPS**
GP	0.0723	0.2900	0.9540	0.0395	0.0872	0.3397	0.9544	0.0482
SPGP	0.0725	0.2900	0.9419	0.0394	0.0872	0.3310	0.9475	0.0480
IMSPGP	0.0654	0.3251	0.9791	0.0356	0.0781	0.3100	0.984	0.0414

Table 3. Comparison of point prediction results in May, August, November, and February.

Month	Model	Sunny	Rainy
		RMSE	**RMSE**
May	IMLSSVM	0.0779	0.0790
	LSSVM	0.0793	0.0823
	IMSPGP	0.0654	0.0781
	QRNN	0.0912	0.0985
August	IMLSSVM	0.1047	0.1275
	LSSVM	0.1070	0.1401
	IMSPGP	0.1040	0.1090
	QRNN	0.1327	0.1210
November	IMLSSVM	0.0933	0.1106
	LSSVM	0.1001	0.1170
	IMSPGP	0.0819	0.1070
	QRNN	0.1075	0.1420
February	IMLSSVM	0.0937	0.1034
	LSSVM	0.1038	0.1172
	IMSPGP	0.0884	0.0966
	QRNN	0.1260	0.1047

Figure 5 shows a scatter plot of the predicted and actual power generation of the IMLSSVM model for the whole year of 2012. As shown in the figure, the data are not evenly distributed. Therefore, when segmenting the predicted power, the number of data in each interval will be different according to the principle of equalization. In order to make the number of samples in each interval roughly the same, the final division of the intervals is shown in Table 4 after several tests.

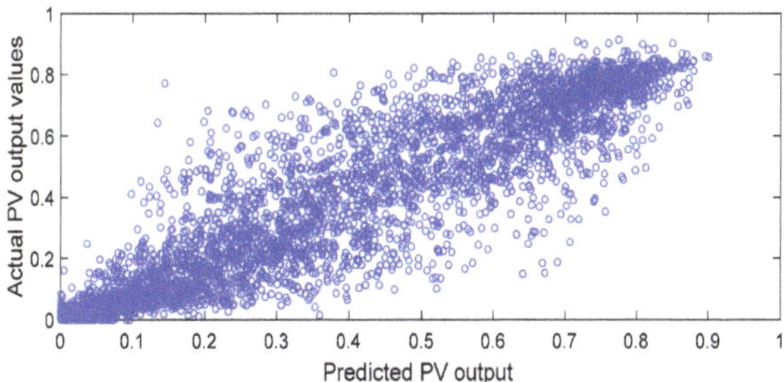

Figure 5. Scatter plot of the predicted and actual values using IMLSSVM.

Table 4. Statistics of samples in each interval.

Interval	[0,0.05]	[0.05,0.2]	[0.2,0.4]	[0.4,0.65]	[0.65,1]
Number of Samples	1268	905	925	951	911

At the first interval, the length is very short, so the range is set to this value. The approximate number of samples in each interval is beneficial to the fitting effect of the parameter estimation, while ensuring the reliability and stability of the error statistics.

In order to find the estimation method that can best describe the model error distribution, this paper uses the normal distribution, T-location-scale distribution, logistic distribution, and KDE method to estimate the probability density of the error values in five intervals. As shown in Figure 6, sub-graphs (a), (b), (c), (d), and (e) represent intervals 1, 2, 3, 4, and 5, respectively. The histogram represents the distribution of errors. It can be seen from the figure that there are different error distributions among different power intervals, so it is feasible to use different probability density functions among different power intervals. In addition, it shows that the probability density curve obtained by the KDE method fits better with error distribution map. Therefore, this paper will use the KDE method to estimate the probability density of different intervals.

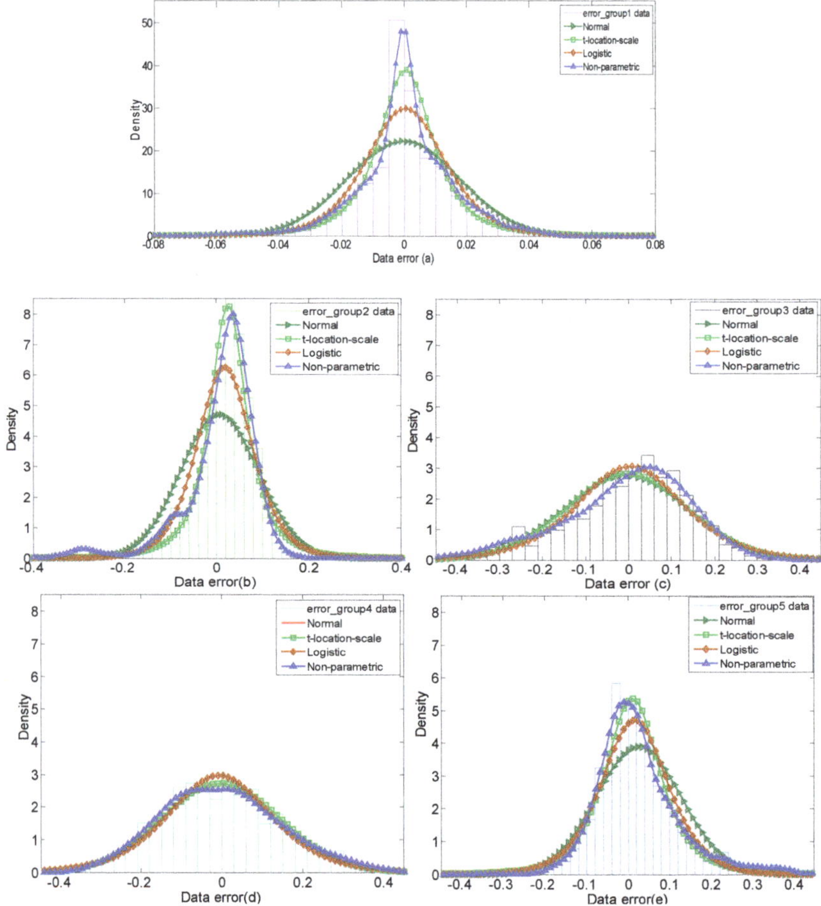

Figure 6. Comparison of error probability density functions in different intervals.

After obtaining the probability density function of each interval, the power interval which the point prediction value of the IMLSSVM belongs to is determined. Then, the probability density function of the interval is used as the probability density function of the prediction point, thereby obtaining the probability distribution. In Figure 7, the forecasting result of seven days in May using IMLSSVM is shown. Compared with Figure 4, the interval width of the IMLSSVM model is similar to the IMSPGP model at the same confidence level. In the high-power interval, the value of width becomes larger, but in the low-power interval, the value of width is slightly smaller than the IMSPGP model, which is related to the peak-tail characteristic of the probability density function in interval I.

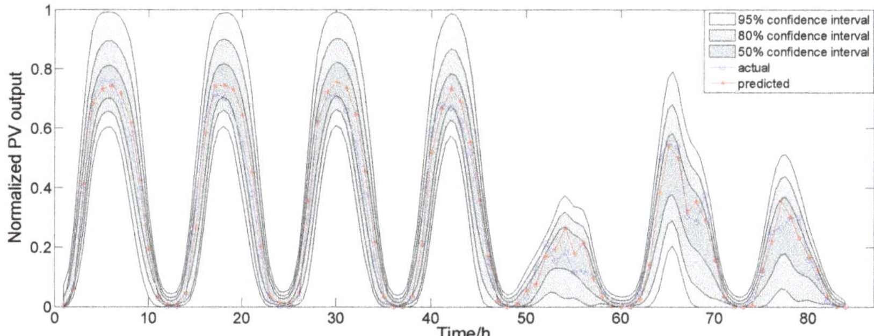

Figure 7. Point forecast and interval forecast over seven days using IMLSSVM.

The cumulative distribution function graph of each model at multiple time points is made to further analyze the probability distribution of the model, as shown in Figure 8. The characteristic of the IMSPGP curve is relatively flat and the performance is general. The QRNN model has steep curve characteristics in different prediction models, indicating that it has the characteristics of sharp peaks and thin tails, and the sharpness features are obvious. As shown in sub-graph (b) and (d), when the power is small, IMLSSVM has increased sharpness compared with other predictive models, and the cumulative distribution function is better. But when the power value becomes large, the predicted cumulative distribution function is not so prominent in many predictive models. It can be known from the above analysis, the three prediction models have different prediction effects.

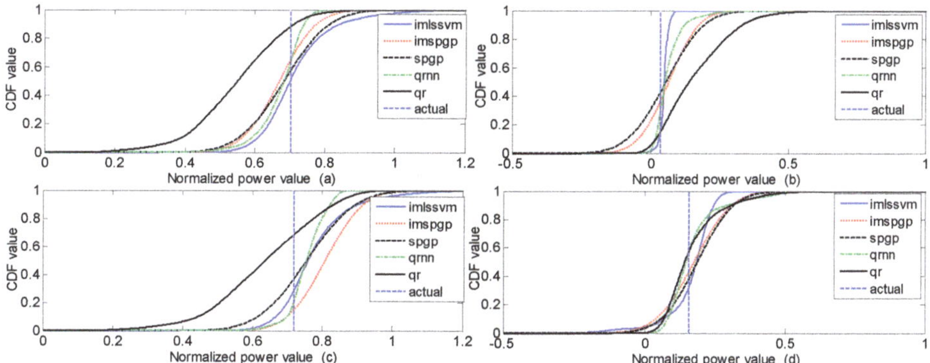

Figure 8. Cumulative distribution function graph.

4.3.2. Analysis of Experiment Results for the Combined Model

In order to obtain a better weight coefficient, this paper uses the arithmetic average method (M1), continuous ranking probability score reciprocal method (M2), IGWO combination method (M3), and entropy method (M4) for comparison and analysis. The CRPS value is taken as the fitness value when the IGWO combination method is used. The number of wolves is set to 30, and the maximum number of iterations is 50. The three weight coefficients of the three single models are variables, and the range are all [0, 1].

Table 5 shows the CRPS of each model in different intervals. It is shown that in the same interval, each mode has a different CRPS. It is feasible to use the combination with variable intervals and weights.

Table 5. CRPS of each model in different intervals.

Interval	Model	Sunny CRPS	Rainy CRPS
I	IMSPGP	0.0176	0.0163
	QRNN	0.0084	0.0042
	IMLSSVM	0.0134	0.0075
II	IMSPGP	0.0520	0.0542
	QRNN	0.0512	0.0618
	IMLSSVM	0.0623	0.0599
III	IMSPGP	0.0414	0.0635
	QRNN	0.0474	0.0672
	IMLSSVM	0.0567	0.0630
IV	IMSPGP	0.0203	0.0228
	QRNN	0.0104	0.0121
	IMLSSVM	0.0161	0.0118

Table 6 shows the weight coefficients of each model obtained by the combination method, where w1, w2, and w3 are the weight coefficient values of IMSPGP, QRNN, and IMLSSVM, respectively. According to the weighting coefficients of each model, the CRPS of various combined models can be obtained, as shown in Table 7. It can be seen from Table 7 that among the four combined methods, the IGWO combination method has the smallest CRPS under different weather types (that is, the probability prediction accuracy is the best among the four methods). Therefore, the three models are combined using the IGWO combination method.

Table 6. Weights of each model in different intervals.

Interval	Weight	Sunny M1	M2	M3	M4	Rainy M1	M2	M3	M4
I	w1	0.3333	0.3413	0.3410	0.3661	0.3333	0.1207	0.0000	0.3407
	w2	0.3333	0.3678	0.6587	0.3094	0.3333	0.5367	1.0000	0.3273
	w3	0.3333	0.2908	0.0003	0.3245	0.3333	0.3425	0.0000	0.3320
II	w1	0.3333	0.3425	0.4330	0.3571	0.3333	0.3387	0.2157	0.3371
	w2	0.3333	0.3626	0.5659	0.3404	0.3333	0.3079	0.1020	0.3269
	w3	0.3333	0.2949	0.0011	0.3024	0.3333	0.3533	0.6823	0.3360
III	w1	0.3333	0.3432	1.0000	0.3593	0.3333	0.3360	0.4695	0.3298
	w2	0.3333	0.3724	0.0000	0.3249	0.3333	0.3418	0.1143	0.3419
	w3	0.3333	0.2844	0.0000	0.3158	0.3333	0.3222	0.4162	0.3283
IV	w1	0.3333	0.2203	0.1010	0.3630	0.3333	0.2241	0.0000	0.3555
	w2	0.3333	0.4191	0.8280	0.2965	0.3333	0.4308	0.4033	0.2998
	w3	0.3333	0.3606	0.071	0.3405	0.3333	0.3451	0.5967	0.3446

Figure 9 is the comparison of point prediction errors over seven days for various prediction models. The RMSE values of the combined model, IMLSSVM model, IMSPGP model, and QRNN model are 0.0380, 0.0468, 0.0439, and 0.0573, respectively. The combined model has the best predictive performance. In Figure 10, subgraph (a), (b), (c), and (d) are cumulative distribution functions for interval I, II, III, and IV, respectively. The CRPS values in the four intervals are shown in Table 8. Referring to Figure 10 and Table 8, the combined model can obtain the lowest CRPS for the predicted points at interval II and IV. In the other two intervals, it is not optimal. That indicates that the combined model does not achieve the optimal CRPS in each interval, but that for the overall test set the combined model has higher probability prediction accuracy than the IMSPGP model, the QRNN model, and the IMLSSVM model, which can be verified in Table 9.

Table 7. CRPS of various combinations.

Model	Sunny	Rainy
	CRPS	CRPS
IMSPGP	0.0348	0.0431
QRNN	0.0326	0.0416
IMLSSVM	0.0407	0.0402
M1	0.0334	0.0395
M2	0.0330	0.0390
M3	0.0306	0.0383
M4	0.0334	0.0395

Table 8. CRPS values in the four intervals.

Model	CRPS			
	I	II	III	IV
Group	0.0339	0.0321	0.0152	0.0247
IMSPGP	0.0306	0.0342	0.0325	0.0257
QRNN	0.0374	0.0393	0.0084	0.0252
IMLSSVM	0.0543	0.0458	0.0198	0.0348

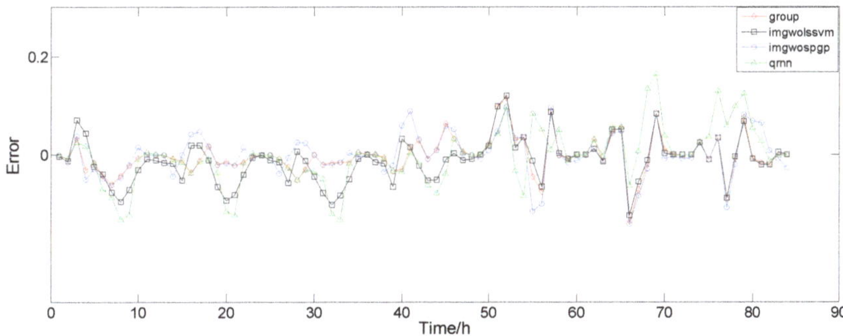

Figure 9. Comparison of point prediction error over seven days.

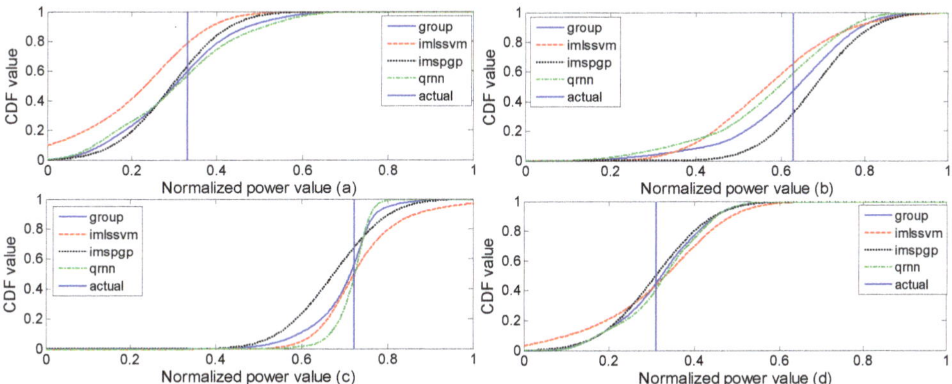

Figure 10. Multi-point cumulative distribution function in different intervals.

Table 9. Evaluation indices of different models during four months.

Month	Model	Sunny				Rainy			
		RMSE	PINAW	PICP	CRPS	RMSE	PINAW	PICP	CRPS
May	IMSPGP	0.0674	0.3160	0.9762	0.0348	0.0856	0.3158	0.9792	0.0431
	QRNN	0.0785	0.3352	0.9345	0.0417	0.1038	0.3108	0.9688	0.0402
	IMLSSVM	0.0961	0.2590	0.9881	0.0326	0.1271	0.3151	0.9688	0.0416
	Group	0.0706	0.2842	0.9762	0.0306	0.0801	0.3129	0.9896	0.0383
August	IMSPGP	0.1077	0.3391	0.9389	0.0538	0.1011	0.3087	0.9583	0.0542
	QRNN	0.1087	0.3698	0.9111	0.0560	0.1250	0.3367	0.9444	0.0560
	IMLSSVM	0.1393	0.4084	0.9611	0.0552	0.1201	0.3861	0.9583	0.0630
	Group	0.1079	0.3405	0.9433	0.0513	0.0988	0.3189	0.9861	0.0480
November	IMSPGP	0.0779	0.3630	0.9703	0.0450	0.1053	0.3085	0.9375	0.0570
	QRNN	0.0937	0.3388	0.9322	0.0491	0.1094	0.3549	0.9375	0.0555
	IMLSSVM	0.0993	0.3339	0.9915	0.0413	0.1210	0.3833	0.1000	0.0487
	Group	0.0808	0.3319	0.9873	0.0397	0.0805	0.3442	0.9792	0.0399
February	IMSPGP	0.0899	0.3241	0.9512	0.0436	0.0876	0.3312	0.9577	0.0489
	QRNN	0.0967	0.3350	0.9329	0.0472	0.0865	0.2940	0.9155	0.0430
	IMLSSVM	0.1142	0.3168	0.9817	0.0423	0.0895	0.3393	0.9718	0.0431
	Group	0.0885	0.3489	0.9878	0.0403	0.0819	0.3420	0.9859	0.0372

In order to further verify the stability and accuracy of the combined model, the data in May, August, November, and February is taken for further verification analysis. The results of each indicator are shown in Table 9. Whether it is sunny or rainy, the CRPS values of the combined model are the smallest. However, on sunny days, there is a case where the IMSPGP model has higher prediction accuracy than the combined model in terms of point prediction. The main reason is that the CRPS is adopted as a fitness value in the IMGWO combination method. Taking the results of May as an example, the combined model has the smallest CRPS value for either sunny days or rainy days. It should be noted that, on sunny days, the PINAW of the IMLSSVM model is smaller than that of the combined model, and the PICP of the IMLSSVM model is larger than the PICP of the combined model. The PINAW and the PICP describe the 95% confidence interval index, which can reflect the probability prediction performance to a certain extent. The CRPS can describe the probability prediction performance comprehensively. According to the comprehensive analysis, the combined model prediction performance improves upon the single model, and has good stability and practicability since the CRPS of the combined model is smaller. A more-accurate probability prediction of the PV power means that uncertainty is reduced, thereby lowering the operating costs of the energy system.

5. Conclusions

In this paper, photovoltaic power generation prediction was deeply studied. Improved sparse Gaussian process and improved least squares support vector machine were proposed, and the accurate results of photovoltaic power generation probability prediction were obtained. Through analysis, we found that the prediction performance of each model was different, and a combined probability prediction model with different interval weights is proposed to make the best of the characteristics of each model. Finally, the simulation results showed that compared with single prediction models, the combined probability prediction model could predict photovoltaic power generation more accurately. Moreover, whether on sunny days or rainy days, high probability prediction accuracy could be guaranteed for different months. Therefore, the model has high stability and practicability.

Author Contributions: Methodology, Q.L. and W.Z.; Supervision, Z.C.; Writing—original draft, Q.L.; Writing—review & editing, W.Z.

Funding: This research was funded by the National Natural Science Foundation of China, grant number 61873180.

Acknowledgments: All the authors thank the International Institute of Electrical and Electronics Engineers Energy Forecasting Working Group for providing the experimental data.

Conflicts of Interest: The authors declare no conflict of interest.

References

1. IRENA. RENEWABLE CAPACITY STATISTICS 2018[EB/OL]. Available online: http://www.irena.org/publications/2018/Mar/Renewable-Capacity-Statistics-2018 (accessed on 18 July 2018).
2. Sobri, S.; Koohi-Kamali, S.; Rahim, N.A. Solar photovoltaic generation forecasting methods: A review. *Energy Convers. Manag.* **2018**, *156*, 459–497. [CrossRef]
3. Singh, G.K. Solar power generation by PV (photovoltaic) technology: A review. *Energy* **2013**, *53*, 1–13. [CrossRef]
4. Junwei, Z. Prediction of Output Power of Photovoltaic Power Plant Based on Physical Model. *Transactions* **2015**, *51*, 46–49.
5. Dolara, A.; Leva, S.; Manzolini, G. Comparison of different physical models for PV power output prediction. *Sol. Energy* **2015**, *119*, 83–99. [CrossRef]
6. Ding, S.; Su, C.; Yu, J. An optimizing BP neural network algorithm based on genetic algorithm. *Artif. Intell. Rev.* **2011**, *36*, 153–162. [CrossRef]
7. Liu, Y.; You, Z.; Cao, L. A novel and quick SVM-based multi-class classifier. *Pattern Recognit.* **2006**, *39*, 2258–2264. [CrossRef]
8. Adankon, M.M.; Cheriet, M. Genetic algorithm-based training for semi-supervised SVM. *Neural Comput. Appl.* **2010**, *19*, 1197–1206. [CrossRef]
9. Johansen, T.A.; FOSS, B. Constructing NARMAX models using ARMAX models. *Int. J. Control* **1993**, *58*, 29. [CrossRef]
10. Hu, F. Markov Chain. *Encycl. Syst. Biol.* **2013**, *14*, 81–93.
11. Chen, C.; Duan, S.; Cai, T.; Liu, B. Online 24-h solar power forecasting based on weather type classification using artificial neural network. *Sol. Energy* **2011**, *85*, 2856–2870. [CrossRef]
12. Chen, J.L.; Liu, H.B.; Wu, W.; Xie, D.T. Estimation of monthly solar radiation from measured temperatures using support vector machines - A case study. *Renew. Energy* **2011**, *36*, 413–420. [CrossRef]
13. Li, Y.; Su, Y.; Shu, L. An ARMAX model for forecasting the power output of a grid connected photovoltaic system. *Renew. Energy* **2014**, *66*, 78–89. [CrossRef]
14. Ding, K.; Feng, L.; Wang, X.; Qin, S.; Mao, J. Forecast of PV power generation based on residual correction of Markov chain. In Proceedings of the International Conference on Control, Bandung, Indonesia, 27–29 August 2015.
15. Eseye, A.T.; Zhang, J.; Zheng, D. Short-term Photovoltaic Solar Power Forecasting Using a Hybrid Wavelet-PSO-SVM Model Based on SCADA and Meteorological Information. *Renew. Energy* **2018**, *118*, 357–367. [CrossRef]
16. Ding, M.; Wang, L.; Bi, R. A short-term prediction model of output power of photovoltaic power generation system based on improved BP neural network. *Power Syst. Prot. Control* **2012**, *40*, 93–99.
17. Yang, H.T.; Huang, C.M.; Huang, Y.C.; Pai, Y.S. A Weather-Based Hybrid Method for 1-Day Ahead Hourly Forecasting of PV Power Output. *IEEE Trans. Sustain. Energy* **2014**, *5*, 917–926. [CrossRef]
18. Hosoda, Y.; Namerikawa, T. Short-term photovoltaic prediction by using H∞ filtering and clustering. In Proceedings of the Sice Conference IEEE, Akita, Japan, 20–23 August 2012.
19. Liu, J.; Fang, W.; Zhang, X.; Yang, C. An Improved Photovoltaic Power Forecasting Model with the Assistance of Aerosol Index Data. *IEEE Trans. Sustain. Energy* **2015**, *6*, 434–442. [CrossRef]
20. Liu, F.; Li, R.; Li, Y.; Yan, R.; Saha, T. Takagi–Sugeno fuzzy model-based approach considering multiple weather factors for the photovoltaic power short-term forecasting. *IET Renew. Power Gener.* **2017**, *11*, 1281–1287. [CrossRef]
21. Zhu, H.; Li, X.; Sun, Q.; Nie, L.; Yao, J.; Zhao, G. A Power Prediction Method for Photovoltaic Power Plant Based on Wavelet Decomposition and Artificial Neural Networks. *Energies* **2016**, *9*, 11. [CrossRef]
22. Malvoni, M.; De Giorgi, M.G.; Congedo, P.M. Photovoltaic forecast based on hybrid PCA–LSSVM using dimensionality reduced data. *Neurocomputing* **2016**, *211*, 72–83. [CrossRef]
23. Marquez, R.; Pedro, H.T.C.; Coimbra, C.F.M. Hybrid solar forecasting method uses satellite imaging and ground telemetry as inputs to ANNs. *Sol. Energy* **2013**, *92*, 176–188. [CrossRef]

24. Xiang, Z.; Rongrong, J. Photovoltaic ultra-short-term power prediction model combining numerical weather prediction and foundation cloud map. *Autom. Electr. Power Syst.* **2015**, *39*, 4–10.

25. Van der Meer, D.W.; Shepero, M.; Svensson, A.; Widén, J.; Munkhammar, J. Probabilistic forecasting of electricity consumption, photovoltaic power generation and net demand of an individual building using Gaussian Processes. *Appl. Energy* **2018**, *213*, 195–207. [CrossRef]

26. Wei-jia, Z.; Ning, Z.; Chongqing, K. Predictive error probability distribution estimation method for photovoltaic power generation. *Autom. Power Syst.* **2015**, *39*, 8–15. (In Chinese)

27. Fatemi, S.A.; Anthony, K.; Matthias, F. Parametric methods for probabilistic forecasting of solar irradiance. *Renew. Energy* **2018**, *129*, 666–676. [CrossRef]

28. Junior, J.G.D.S.F.; Oozeki, T.; Ohtake, H.; Takashima, T.; Kazuhiko, O. On the Use of Maximum Likelihood and Input Data Similarity to Obtain Prediction Intervals for Forecasts of Photovoltaic Power Generation. *J. Electr. Eng. Technol.* **2015**, *10*, 1342–1348. [CrossRef]

29. Almeida, M.P.; Perpinan, O.; Narvarte, L. PV power forecast using a nonparametric PV model. *Sol. Energy* **2015**, *115*, 354–368. [CrossRef]

30. Sanjari, M.J.; Gooi, H.B. Probabilistic Forecast of PV Power Generation Based on Higher-Order Markov Chain. *IEEE Trans Power Syst.* **2017**, *32*, 2942–2952. [CrossRef]

31. Yang, Y.; Li, S.; Li, W.; Qu, M. Power load probability density forecasting using Gaussian process quantile regression. *Appl. Energy* **2018**, *213*, 499–509. [CrossRef]

32. Edward, S.; Ghahramani, Z. Sparse Gaussian Processes using Pseudo-inputs. In Proceedings of the International Conference on Neural Information Processing Systems, Vancouver, BC, Canada, 3 June 2010.

33. Mirjalili, S.; Mirjalili, S.M.; Lewis, A. Grey Wolf Optimizer. *Adv. Eng. Softw.* **2014**, *69*, 46–61. [CrossRef]

34. Suykens, J.A.K.; Vandewalle, J. Least Squares Support Vector Machine Classifiers. *Neural Process. Lett.* **1999**, *9*, 293–300. [CrossRef]

35. Malec, P.; Schienle, M. Nonparametric kernel density estimation near the boundary. *Comput. Stat. Data Anal.* **2014**, *72*, 57–76. [CrossRef]

36. Yu, C. *Research and Application of Combined Forecasting Model*; Shandong University: Jinan, China, 2017.

37. Ze, C.; Chong, L.; Li, L. Estimation Method of Photovoltaic Output Probability Distribution Based on Similar Time. *Power Syst. Technol.* **2017**, *41*, 448–454.

38. Huayou, C. *The Effectiveness Theory of Combined Forecasting Methods and its Application*; Beijing Science Press: Beijing, China, 2008; pp. 20–87.

39. Global energy forecasting competition 2014[EB/OL]. Available online: http://www.crowdanalytix.com/contests/global-energy-forecasting-competition-2014-probabilistic-solar-power-forecasting (accessed on 29 May 2017).

40. Archer, K.J.; Kimes, R.V. Empirical characterization of random forest variable importance measures. *Comput. Stat. Data Anal.* **2008**, *52*, 2249–2260. [CrossRef]

41. Gneiting, T.; Balabdaoui, F.; Raftery, A.A.E. Probabilistic forecasts, calibration and sharpness. *J. R. Stat. Soc.* **2007**, *69*, 243–268. [CrossRef]

Article

Is Deployment of Charging Station the Barrier to Electric Vehicle Fleet Development in EU Urban Areas? An Analytical Assessment Model for Large-Scale Municipality-Level EV Charging Infrastructures

Giacomo Talluri [1,2], **Francesco Grasso** [1] and **David Chiaramonti** [2,3,*]

[1] Department of Information Engineering, University of Florence, Via di S. Marta 3, 50139 Florence, Italy; giacomo.talluri@unifi.it (G.T.); francesco.grasso@unifi.it (F.G.)
[2] RE-CORD, Viale Kennedy 182, Scarperia e San Piero, 50038 Florence, Italy
[3] Department of Industrial Engineering, University of Florence, Viale Morgagni 40, 50135 Florence, Italy
* Correspondence: david.chiaramonti@re-cord.org; Tel.: +39-055-275-8690

Received: 15 September 2019; Accepted: 29 October 2019; Published: 4 November 2019

Abstract: This work investigates minimum charging infrastructure size and cost for two typical EU urban areas and given passenger car electric vehicle (EV) fleets. Published forecasts sources were analyzed and compared with actual EU renewal fleet rate, deriving realistic EV growth figures. An analytical model, accounting for battery electric vehicle-plug-in hybrid electric vehicle (BEV-PHEV) fleets and publicly accessible and private residential charging stations (CS) were developed, with a novel data sorting method and EV fleet forecasts. Through a discrete-time Markov chain, the average daily distribution of charging events and related energy demand were estimated. The model was applied to simulated Florence and Bruxelles scenarios between 2020 and 2030, with a 1-year timestep resolution and a multiple scenario approach. EV fleet at 2030 ranged from 2.3% to 17.8% of total fleet for Florence, 4.6% to 16.5% for Bruxelles. Up to 2053 CS could be deployed in Florence and 5537 CS in Bruxelles, at estimated costs of ~8.3 and 21.4 M€ respectively. Maximum energy demand of 130 and 400 MWh was calculated for Florence and Bruxelles (10.3 MW and 31.7 MW respectively). The analysis shows some policy implications, especially as regards the distribution of fast vs. slow/medium CS, and the associated costs. The critical barrier for CS development in the two urban areas is thus likely to become the time needed to install CS in the urban context, rather than the related additional electric power and costs.

Keywords: electric vehicles EV; optimal sizing; charging infrastructure; Markov chain; EV fleet forecasts; decarbonization

---

1. Introduction

Mitigating the effects of climate change through greenhouse gas (GHG) emissions reductions is one of the key challenges of the 21st century. At the core is the issue of overall energy consumption as well as the need for stronger decarbonization policies. Within this picture, transport sector plays a big role, globally and at European level: EEA data shows that in 2016, it accounted for a 33.2% share of EU-28 final energy consumptions and for a 24.3% share of GHG emissions [1]. Road transport accounted for 72% of GHG emissions and, within that sector, cars accounted for 60.7%. Moreover, half of EU-28 NO_x emissions and at least 15% of PM10, PM2.5, SO_x and CO emissions are transport-related [1,2]; finally, European transport energy needs are fulfilled by fossil fuels use for more than a 94% share [1].

Several alternative energy sources and renewable fuels are available—with different level of technology readiness and market penetration—in order to reduce dependence on fossil fuels: among them we find biofuels such as biodiesel and bioethanol, renewable hydrocarbons, ligno-cellulosic ethanol, biomethane, renewable fuels of non-biological origin (RFNBO) recycled carbon fuels, renewable e-fuels, renewable hydrogen, electricity, etc. None of these alternative fuels and sources alone will ever completely substitute fossil fuels, at least in a short term, and most of the forecasts see them coexisting in future fuel mixes [3]; in 2016, in EU-28, biofuels accounted for the bigger share, covering 4.6% of total final fuel consumption [1].

Anyway, rapid cost reduction of solar and wind power technologies created strong prospects for further electrification of end-use sectors. Within this framework, electric vehicles (EVs)—especially passenger cars—are one of the most pursued solutions to achieve large-scale transport sector decarbonization; anyway, they are not a "drop-in" solution, since they need a new and alternative charging infrastructure. Official documents such as the Alternative Fuels Infrastructure (AFI) Directive [4] and the National Plans for alternative fuels and EV charging infrastructure such as [5,6] are of interest because they express governments forecasts about future EV fleet size and formal recommendations on topics such as the minimum number of charging stations (CS) to be deployed and the minimum EV/CS expected ratio.

It is expected that urban areas will be the place of a large-scale infrastructure deployment in short to mid-term, since they are the most suitable for actual EV use and would receive maximum benefit from noise and local polluting emission reductions that are related with the shift from internal combustion engine vehicles (ICEV) to EV. To this regards, stakeholders and decision makers could benefit from reliable predictions on the dimension and cost of a charging infrastructure suitable for, i.e., a municipality area.

Literature already presents several studies on EV charging infrastructures, which analyze the topic from a multitude of points of view. Many of them evaluate the optimal positioning of a given set of CS through geographic information system (GIS) procedures, such as [7,8] or through traffic flows analysis such as [9]; they develop a model taking into account residential statistics, parking area information, electric power distribution network position and other data to define the optimal position for charging stations, but usually the input data about the number of CS to be deployed have little or no connection with data on EV fleets circulating in the area or on forecasts about that. Other studies focus on business and profitability analysis of a specific CS installation [10], while other analyze the possible interactions between charging EV and RES electrical generation that could take place in urban areas, such as the one from PV installations [11,12].

Finally, several papers analyze the topic of optimal sizing of EV charging infrastructures. Unfortunately, reports that specifically evaluate cost and size of large-scale deployments, using real data and forecasts (such as [13,14]) usually refer to national or international levels and give as output highly aggregated information; this makes it difficult, afterwards, to scale them down to a more local level. Several studies can be found, that address the problem at a regional and municipality level: [15] analyzes the German region of Stuttgart, but it focuses its model on CS availability rate and is not applied to a real scenario. The study [16] is related to the Italian province of Florence and uses GPS real data from around 12,000 ICE vehicles, extracting driving and parking patterns to size the charging infrastructure; anyway, it does not explicitly give as an output a number of CS and the relative cost, and it does not consider a temporal evolution of the analyzed situation. Lastly, [17,18] give as an output the forecast size of the charging infrastructure for two different Swiss municipalities, but neither of them give a proper explanation on the methodology used.

Scope of this work is to define the optimal size of a minimum cost charging infrastructure, suitable for deployment in an urban area and able to cope with the requests of a given passenger car EV fleet. This study focuses only on EV passenger cars since they account for the vast majority of EU-28 circulating road vehicles with an 87% share of total, while, in comparison, light commercial vehicles accounted for 9.8% (authors elaboration on 2016 Eurostat data [19]). Moreover, this paper will investigate the impact of such infrastructure on the existing parking stalls and on the electrical network, in terms of the average energy and power requests; finally, it will characterize the infrastructure using performance indicators (PI) such as average daily charging events and global utilization rate.

In order to address these requests, an analytical model has been developed, separately accounting for battery electric vehicle (BEV) and plug-in hybrid electric vehicle (PHEV) fleets, as well as for publicly accessible and private residential CS.

Given the primary importance of the inputs related to the size of the circulating EV fleet, an extensive literature research for EV fleet growth forecasts has been carried out, focusing on reports related to the two considered areas as well as to the analyzed timeframe. Since substantial differences between the various forecasts emerged from the research, a novel method for data sorting and conditioning has been developed, using circulating fleet turnover rate as threshold indicator.

The model has then been applied to the two geographical areas of Florence Municipality for Italy and Bruxelles for Belgium, over the 2020–2030 period, with a 1-year timestep resolution. Within each area nine scenarios have been evaluated; they are obtained as the combination of three different EV circulating fleet forecasts with three possible user's charging behaviors. The results from this study could be used to provide further insights to policy makers and local authorities, in order to better understand and manage the transition period toward a higher share of urban electrical mobility.

To the best of the authors' knowledge, within the available studies, the one which shares the closest similarities with present work's purposes is [20]; anyway, it describes the charging infrastructure with a lower level of details and it does not implement a model for the estimating of EV fleet growth, using as input data coming from a single source.

This research article is structured as follows: in Section 2, the inputs needed by the model together with the related procedures are first defined, whenever relevant; then, the model structure and embedded algorithms are thoroughly presented. Finally, the outputs and the rationale for the input scenarios implemented in this paper are reported. The first part of Section 3 reports details of the numerical values of the inputs used in the various scenario, while the second part discusses the obtained results, using also several PI. Finally, summary conclusions and closing remarks are given in Section 4.

2. Model and Methodology

2.1. Model Definition

In order to define the optimal size of the EV charging infrastructure, the model needs as inputs, for each timestep of the analysis, the number of daily charging events related to both BEV and PHEV fleets and the corresponding energy requests; moreover, it needs a full characterization of the CS in terms of expected performances and costs. An input pre-processing phase take place before the model's core steps execution, and a post-processing phase, delivering expected outputs is placed after them; this leads to a four-parts conceptual architecture, thoroughly described in the following sections and also briefly reported below and in Figure 1:

- The first part is dedicated to data collection and pre-processing.
- The second part performs a discrete-time Markov process to obtain as output the probability distribution of the charging events over the average range (in days) of BEVs and PHEVs; to do so it also considers users' behavior and preferences. Several control criteria are implemented, in order to assure coherence of the results.
- In the third part the outputs are used to define a constrained space of solutions where all the possible charging infrastructures complying with EV fleet requests are defined; finally, CS costs are applied, and the optimal charging infrastructure size and composition is chosen.
- In the fourth and last part the results are processed in order to obtain the desired outputs.

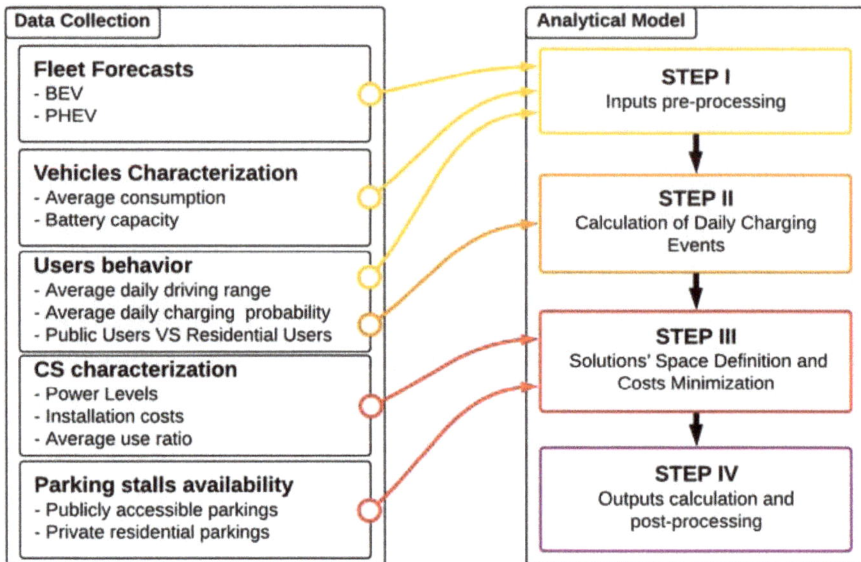

Figure 1. High-level visual description of the model structure.

The model structure has been conceived so to be flexible enough to capture the complexity of the evaluated scenarios, while maintaining a simple and lean structure. These initial requirements led to the following set of features:

- Modular architecture and additivity of the model: Each timestep of the analysis is considered separately and the model is recursively operated. This approach is used as well within a single timestep: as an example, BEV and PHEV impacts on infrastructure are separately calculated using the same module. Partial outputs are collected at the end of each calculation step of and then post-processed in a final step, in order to obtain the total outputs. This feature allows us to model flexible and time-evolving scenarios, while remaining sufficiently simple. The downside of this choice is that it does not allow any change in the operating parameters of the already deployed infrastructure during the timespan of the analysis.

- Complete battery charge at every charging event: This simplifying assumption derives from the fact that only few information is available on this specific charging behavior; moreover, it does not affect the average quantity of energy requested from the charging infrastructure, since it only depends on average EV fleet size and consumption.

- All vehicles are used every day: This simplifying assumption set vehicles usage pattern as constant over time. This, together with the evaluation through the average daily travelled distance, allows us to define an average usage of the charging infrastructure.

- Evaluation of both BEV and PHEV impacts on charging infrastructure: Given the difference in battery capacities, thus in energy requests during the charging phase, the model takes them into account separately. This choice allows us to better evaluate the impact on charging infrastructure, thus, to dimension it more precisely.

- Implementation of publicly accessible and private residential charging infrastructure: Three different CS power levels are considered within the model to define publicly accessible charging infrastructure; the CS types used can also be changed through the evaluation time period. Moreover, it is of primary importance to investigate also the possible extension and impact of a residential charging infrastructure, since a high number of CS connected to residential distribution feeders could possibly lead to line congestions and voltage issues [21]. In fact, several studies state that early EV adopters are likely to own garages or parking spaces [22] and, more generally, 25–40% (depending on EU countries) of vehicles owners are also garages or parking owners [20,23].

2.2. Model Structure

2.2.1. Data Collection

This section describes all the necessary inputs, as well as the data collection methods, while Section 3.1 reports the specific values of the actual data used during model implementation, together with a description of the data sources. The data collected can be classified under five macro-categories, spanning from BEV and PHEV fleets size forecast, to their average range and consumption; from parking spaces availability to users' behaviors and finally to CS characteristics and costs.

BEV and PHEV fleets forecast over the analysis timeframe:

As already specified in the introduction, only the M1 category light passenger car sector would be considered in this paper. According to [24], M1 category vehicles are designed and constructed for the carriage of passengers, comprising no more than eight seats in addition to the driver's seat. A literature research was then carried out for forecasts on BEV and PHEV car fleets size, over the 2020–2030 period, with a specific focus in years 2020, 2025 and 2030 and referring to the two evaluated areas. Unfortunately, no municipality-level results were publicly available, thus the research was further extended to forecasts evaluating single EU countries, as well as the EU-28 region as a whole. Following this decision, a methodology to scale down National and EU-28-level data back to the context of a municipality was then developed. In order to check dataset soundness for model's scopes, the methodology also compares the collected forecasts to the existing market conditions, using the EU-28 global passenger car fleet turnover rate as a threshold for xEV fleet forecast growth rate.

Being the xEV market relatively new and still evolving, it presents different penetration levels across EU countries; this situation can be related to local factors such as the current development of the charging infrastructures and the existence (or the lack) of active support schemes and subsidies. On the other hand, the total passenger vehicle market—mainly composed by ICEV—is well-developed and with a relatively stable trend in terms of number of circulating vehicles. Following the previous considerations, the methodology has been developed under two main assumptions:

- Within 2050, the xEV distribution across EU-28 countries will follow that of total passenger fleet.
- The forecasts downscaling is realized sequentially: from EU-28 level to national level, then from the national level to municipality level.

As the first step of the process, a baseline of auxiliary information is defined for each of the three geographical levels, to be used by the transfer formulas during dataset downscaling. It is composed by an historical dataset, evaluated over the 2012–2017 period and composed by four specific entries:

- Total circulating fleet (ICEV, BEV and PHEV): TOT;
- Total circulating xEV (BEV and PHEV): xEV;
- Total new vehicles registrations (ICEV, BEV and PHEV): NR_{TOT};
- Total new xEV registrations: NR_{EV}.

The last two variables, namely new vehicles and new xEV registrations, were selected as control parameter to check the soundness of the total circulating fleet and total circulating xEV data. Moreover, the auxiliary baseline comprises a forecast value of 2050 total circulating fleet (TOT), obtained by literature research.

After this preparation phase, the downscaling process of the xEV fleet forecasts was performed with the following procedure (the 'input' subscript refers to the geographical area analyzed by the forecast; the 'output' subscript refers to the geographical area to which the forecast was being scaled to):

- At the starting year, i.e., 2017, the historical value of $\left(\frac{xEV_{output}}{xEV_{input}}\right)_{2017}$, which expresses the ratio between the xEV fleet circulating in the two considered areas (i.e., EU-28 and Italy), was assumed.

- At 2050, $\left(\frac{xEV_{output}}{xEV_{input}}\right)_{2050}$ was assumed equal to $\left(\frac{TOT_{output}}{TOT_{input}}\right)_{2050}$, which expresses the ratio between the total circulating car fleets forecast in the two considered areas.

- Finally, $\left(\frac{xEV_{output}}{xEV_{input}}\right)_Y$ was related to the Y-th year of the considered time period and was calculated under the assumption of a linear behavior, as described in Figure 2.

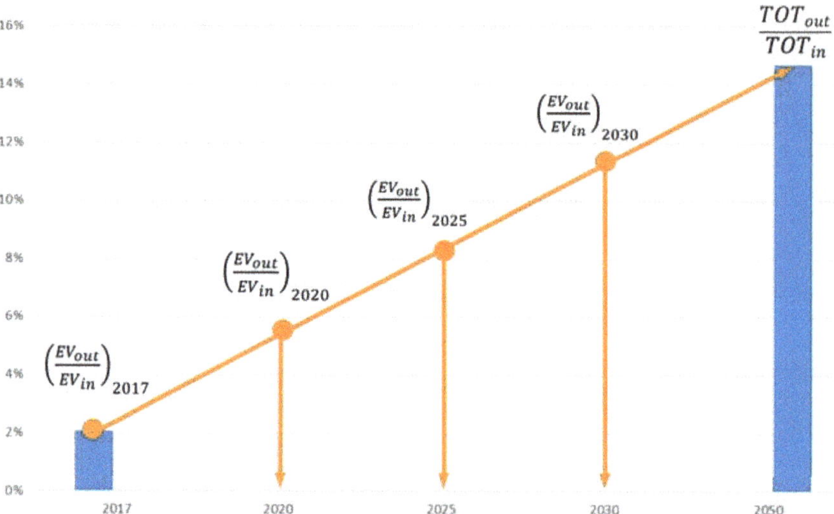

Figure 2. Scaling down process for the collected xEV fleet size forecast values.

Once the forecasts for the xEV fleets size were scaled down to the municipality level, the resulting values were compared, for every year of the period, with a threshold value Th_y, calculated as following:

$$Th_y = \sum_{i=1}^{y} [To * (share_{xEV})_i * TOT_i], \tag{1}$$

where To is the average EU-28 car fleet yearly turnover, with a value of 5.4% [25,26] and $(share_{xEV})_i$ is defined as the xEV share of the EU-28 yearly car turnover, variable over 2020–2030 the period and assuming the values shown in Table 1. Currently, the share of xEV in the annual turnover equals 0.24% for Italy (2017 data) and stays below 4% for most of the European countries [13]; however, it has to be considered the impact that incentives and policies may have on the development of the EV market and the high probability of being activated in the period considered by this study, as highlighted by [13].

Table 1. Assumed xEV share of EU-28 yearly car turnover for the various years of the analyzed timeframe.

	2020	2025	2030
xEV share of EU-28 yearly car turnover	5%	15%	50%

Finally, TOT_y defines the forecast total circulating fleet (ICEV, BEV and PHEV) on the y-th year of period.

A specific forecast was used in the following steps only if all its values were below the threshold, otherwise it was discarded. The equation used for evaluation is described below:

$$xEV_y \leq Th_y \forall y \in (time\ period). \tag{2}$$

Finally, the remaining municipality level forecasts were used to define three scenarios for each municipality considered, using the following criteria:

- Low Scenario: it uses the lowest value of all the selected forecasts for every year of the time period.
- Medium Scenario: it uses an average value calculated from the values of all the selected forecasts for every year of the time period.
- High Scenario: it uses the highest value of all the selected forecasts for every year of the time period.

Average BEV and PHEV energy consumption and batteries capacity:

Average xEV consumption (expressed in kWh km^{-1}) and battery capacity (expressed in kWh) has been considered as a variable, to reflect the inevitable technological advances that will take place over the analyzed period. The values attributed for the year 2020 were obtained from the analysis of the current BEV and PHEV fleet [27] average consumption—measured using WLTP cycle estimations [28]—and capacity of the battery pack.

More specifically, the capacity of the battery pack assigned for the first year of analysis timeframe has been defined as the average of the capacities of the 15 best-seller M1-class BEV (and PHEV), weighted by sales volumes [29]. In order to consider the actual battery discharging capacity the obtained values have been reduced by 30% [28], then assigned to variables $(batt_{PHEV})_{2020}$ and $(batt_{BEV})_{2020}$.

The same approach was also used to define the average mean BEV consumption $(avc_{BEV})_{2020}$, while to obtain the average mean PHEV consumption $(avc_{PHEV})_{2020}$ a further step was required, since their typical use involves the simultaneous operation of both thermal and electric motor. A literature study shown that on average PHEVs cover about 32%–55% of their mileage using electric energy taken from the grid [30]; another approach, presented in [31], suggests an electric load reduction of about 50% compared to BEV vehicles. With a view to make the model easier to implement, the approach of [31] was chosen to define the $(avc_{PHEV})_{2020}$ value, thus defined as the half of $(avc_{BEV})_{2020}$.

On the upper end of the timeframe, namely 2030, these variables were estimated through literature research ([32,33] for consumption and [31,34] for battery capacities). The figures for the remaining years of the period were obtained as linear interpolation of the two extremes.

Public and private stalls availability:

The objective was to assess the availability of adequate space for the installation of private and publicly accessible CS, with a view to define upper limits to the planned charging infrastructure and to allow the evaluation of its impact on stalls occupation.

The following input variables were defined: public parking stalls available on Florence and Bruxelles areas PP_{FI}, PP_{BXL} and private residential parking available on Florence and Bruxelles areas PR_{FI}, PR_{BXL}. Data has been collected through research on an existing database; all these variables are defined as constant during the evaluated period.

CS characteristics definition:

The proper functioning of the model requires the characterization of the charging infrastructure in terms of:

- Charging power levels $(P_k)_y$: the number of power levels and the related power outputs are defined after literature research; private residential and publicly accessible CS are accounted separately.
- Capital costs of the various types of CS $(C_k)_y$: they take into account CS cost, installation and grid connection costs; operation and maintenance costs are not considered.
- Estimated utilization rate of the various types of CS $(r_{k,j})_y$: it is expressed in terms of the maximum number of charging events manageable by the CS on an average daily basis. They depend on the assumed daily usage timeframe h_k, on the charging power level $(P_k)_y$ and on the energy request

of the charging event E_j; subscript k relates to the power level, while subscript j relates to the charging energy request class. They are calculated by comparing the hours of assumed daily availability with the time needed for a charge:

$$\left(r_{k,j}\right) = \frac{h_k}{\left(\frac{E_j}{(P_k)_y}\right)}. \tag{3}$$

The values assumed by these three variables can be updated during the analyzed period; within this study the same values were used for each analyzed municipality.

Driving and charging behavior of users:

The driving and parking habits of users, together with the way they are expected to use the EV charging infrastructure, have a great impact on its characterization, and thus on model outputs. A set of three variables was implemented in order to describe this scenario:

- Average daily driven distance: A literature research focused on urban areas did not return appropriate results, so national level values were used. Anyway, given the fact that urban travels are shorter than the average, the data used was more conservative in terms of energy request to the infrastructure. The values were considered as constant over the timeframe, but differentiated within the two considered areas: avd_{FI} and avd_{BXL}.

- Use of publicly accessible or private residential CS: It is crucial to estimate the share of BEV and PHEV that will weigh on average on the public charging infrastructure; therefore, a literature research has been carried out in order to estimate the percentage of BEV and PHEV that will use the public charging infrastructure over the y-th year of the period: $(\%BEV_P)_y$ and $(\%PHEV_P)_y$.

- Charging events probability distribution over the estimated range of the vehicle: Usually xEV driving range allows for more than one day of use so owners can decide to charge their vehicles when state of charge (SOC) approaches the minimum level or before. This consideration, together with the hypothesis of only complete recharges, leads to different possible energy requests for the single charging event. Thus, it was necessary to develop a methodology to distribute the probability of the charging events over the whole driving range allowed by the battery size. Since this situation is strongly related to user behavior modeling, in order to cover the various possibilities, three different scenarios have been developed, each with a specific probability distribution over the timeframe. The independent variable is represented by the time elapsed since the last charging event (normalized to the maximum autonomy) and is therefore included in an interval [0,1]; the dependent variable p is the probability of a charge event at a given time, defined as a monotonous increasing function with its values included in the interval [0,1]. Immediately after a charging event $p = 0$, while at the end of the driving range $p = 1$, thus avoiding the possibility that a xEV runs out of charge. Figure 3 shows the trends of the three functions defining the different scenarios.

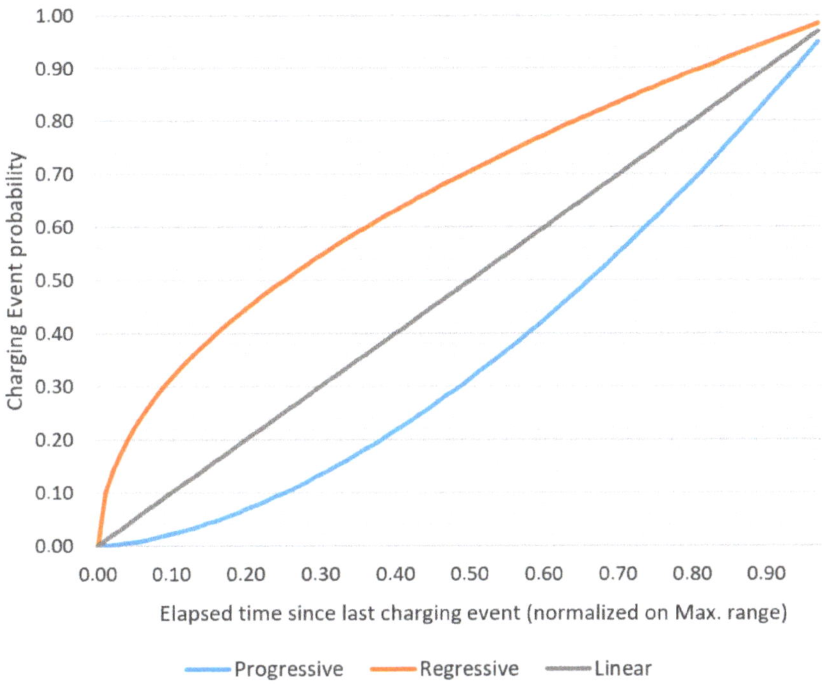

Figure 3. Scenarios for charging events probability distribution.

2.2.2. I Step: Input Pre-Processing

Some of the inputs described in the previous section have direct use in the model, while others need to be further processed. More specifically, the forecasts on BEV and PHEV fleets sizes through the analyzed timeframe were collected as inputs; the model will evaluate separately the number of BEV and PHEV charging using publicly accessible infrastructure from the ones that will use private residential infrastructure, so the following new variables have to be defined:

$$\begin{cases} (BEV_P)_y = (\%BEV_P)_y * (BEV)_y \\ (BEV_R)_y = \left[1 - (\%BEV_P)_y\right] * (BEV)_y \\ (PHEV_P)_y = (\%PHEV_P)_y * (PHEV)_y \\ (PHEV_R)_y = \left[1 - (\%PHEV_P)_y\right] * (PHEV)_y \end{cases}, \tag{4}$$

where $(BEV_P)_y$ and $(PHEV_P)_y$ refers to the shares charging on public infrastructure and $(BEV_R)_y$ and $(PHEV_R)_y$ refers to the shares charging on private residential infrastructure, both over the y-th year of the time period.

Finally, in order to define xEV range as the maximum allowable period of time (measured in days) between two successive charging events, it has to be defined the new variable $[(avr_V)_A]_y$ as:

$$[(avr_V)_A]_y = \frac{(batt_V)_y}{(avc_V)_y * avd_A}, \tag{5}$$

where subscript A refers to the geographical area, V to the type of vehicle (BEV or PHEV) and y to the y-th year of the timeframe. This general notation will be used in the following sections of this document.

2.2.3. II Step: Calculation of Daily Charging Events on Publicly Available Infrastructure

In this step of the model the average number of daily charging events related to the BEV and PHEV circulating fleet is calculated, together with the corresponding energy requests. To do so, the average number of daily charging vehicles and their SOC at the beginning of the charge have to be estimated. BEV and PHEV have different electric driving ranges, thus they are separately evaluated within the model; anyway, since the operations are conceptually identical, in the followings of this section we would simply refer to xEV without losing generality. A discrete-time Markov chain (DTMC), applied to a countable, finite state-space was used to obtain the estimate of the average daily distribution of the SOC levels within the xEV circulating fleet. The DTMC is defined by the transition matrix $[A]$, which values are obtained by applying the charging events probability distribution p over the specific driving range considered, with a 1-day timestep:

$$
[A] = \begin{pmatrix}
p_1 & (1-p_1) & 0 & \cdots & 0 \\
p_2 & 0 & (1-p_2) & \cdots & 0 \\
\vdots & \vdots & \vdots & \ddots & \vdots \\
p_{n-1} & 0 & 0 & \cdots & (1-p_{n-1}) \\
p_n = 1 & 0 & 0 & \cdots & 0
\end{pmatrix}.
\tag{6}
$$

The transition matrix defines the charging probability of a xEV for each day of the driving range; once the driving range $[(avr_V)_A]_y$ is defined, also the dimension of the state-space and of the transition matrix $[A]$ are set accordingly. Figure 4 shows an example of DTMC graph, applied to a 4-days driving range scenario.

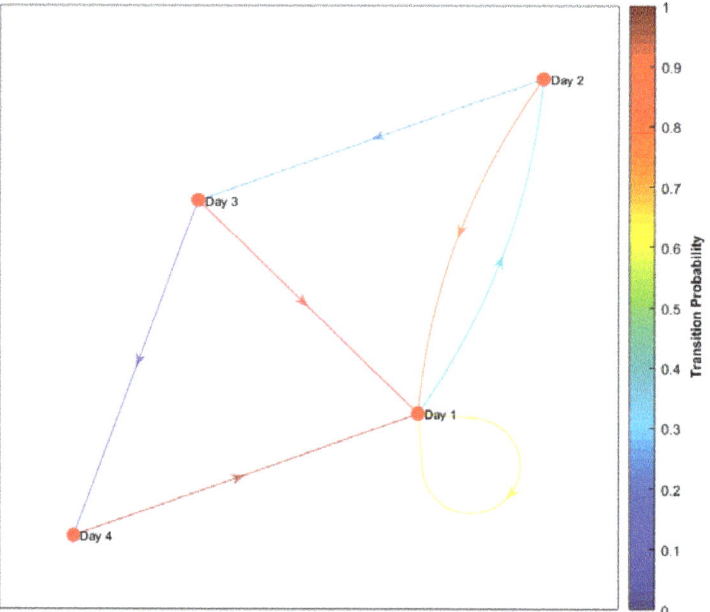

Figure 4. Discrete-time Markov chain (DTMC) plot example, showing the transition probability over a 4-days driving range scenario.

The evolution of the DTMC is calculated with an iterative process—highlighted by Figure 5—that usually reached convergence within 15 steps. The final stationary probability distribution describes the average distribution of SOC levels within the xEV circulating fleet. The application of the transition matrix $[A]$ to this stationary distribution, finally allows us to stochastically define the average number of xEV daily charging, for each SOC level.

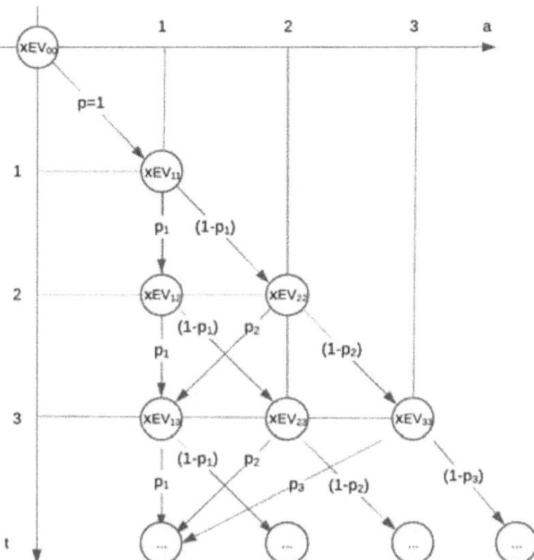

Figure 5. DTMC iterative process graph, showing the probability distribution across driving range days (axis "a") and time (days, axis "t").

The horizontal axis represents timesteps "a", which are inscribed within the interval $\left[0, [(avr_V)_A]_Y\right]$, while the vertical axis represents the iteration steps. A different SOC is associated to every timestep "a", thus a different charging energy request; taking into account the hypothesis of only complete charges its value E_a can be calculated as:

$$[(E_a)_V]_y = [(avc_V)_A * avd_A]_y * a. \tag{7}$$

The output is then a series of couples (xEV_a, E_a), dividing the charging xEV in different groups in terms of energy requests.

In order to check the stability of the iterative process and the quality of the results, two control methods were implemented:

- The sum of the elements $a_{i,j}$ for each row of the transition matrix $[A]$ must be equal to one, since they define all the possible transition events completely:

$$\sum_{j=1}^{j} a_{i,j} = 1. \tag{8}$$

- The total energy recharged by infrastructure (after convergence) must be equal to the average energy consumed by the xEV fleet using public charging infrastructure (Figure 6):

$$[(xEV_P)_A]_y * [(avc_V)_A * avd_A]_y = \sum_{a=1}^{[(am_{xEV})_A]_y} xEV_{a,t} * E_a, \; \forall t > \bar{t}, \tag{9}$$

where $\bar{t}$ is the convergence iteration step.

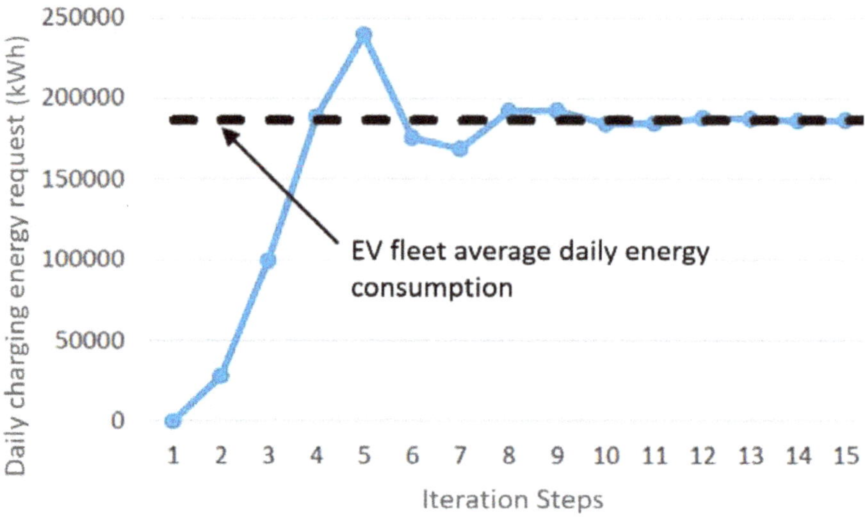

Figure 6. Sample comparison of total energy recharged by infrastructure vs. the average energy consumed by xEV circulating fleet.

2.2.4. III Step: Publicly Available Infrastructure Solutions' Space Definition and Costs Minimization

Single event charging energy requests E_a are related to specific consumption $[(avc_V)_A]_y$ and to battery capacity $(batt_V)_y$, thus they are different between BEV and PHEV; they also evolve during the time period. The overall range of variation of E_a spans from zero to the maximum value of battery capacity $(batt_{BEV})_{2030}$; in order to simplify the model structure, this range has been divided in 10 equally spaced classes, each one with a constant energy value E_j, with:

$$\begin{cases} E_j = j * \frac{(E_a)_{max}}{10} \\ E_0 = 0 \end{cases}, 1 \leq j \leq 10. \tag{10}$$

Then, every E_a value has been compared with E_j, so that for every $E_{j-1} \leq E_a \leq E_j$ the model apply the substitution $E_a = E_j$. This way, several different values of energy requests E_a are reduced to only ten values of E_j; this assumption is safe since it always overestimates the energy requests. The couples (BEV_a, E_a) and $(PHEV_b, E_b)$ are transformed into $\left((BEV_j + PHEV_j), E_j\right)$, thus inscribing BEV and PHEV energy requests into the same framework and allowing us to use the additive and modular architecture that was one of the basic choices for the model.

Given the modular architecture of the model, infrastructure size is calculated for every j-th class of energy requests and, after the cost optimization phase, the total value is obtained by summation. Since $\left(r_{k,j}\right)_y$ already takes into account charging energy requests, all the public infrastructure compositions that are able to satisfy the total number of daily charging events $R_j = \left(BEV_j + PHEV_j\right)$ are considered as suitable. The space of the solutions, for every j-th class, is a triangular portion of a plane, as Figure 7 shows, described by the following equations:

$$\begin{cases} R_j = r_{s,j} * s_j + r_{m,j} * m_j + r_{f,j} * f_j \\ s_j \geq 0 \\ m_j \geq 0 \\ f_j \geq 0 \end{cases}, \tag{11}$$

where the triplets $\left(s_j, m_j, f_j\right)_n$ define all the n possible combinations of CS that satisfy total daily charging requests R_j.

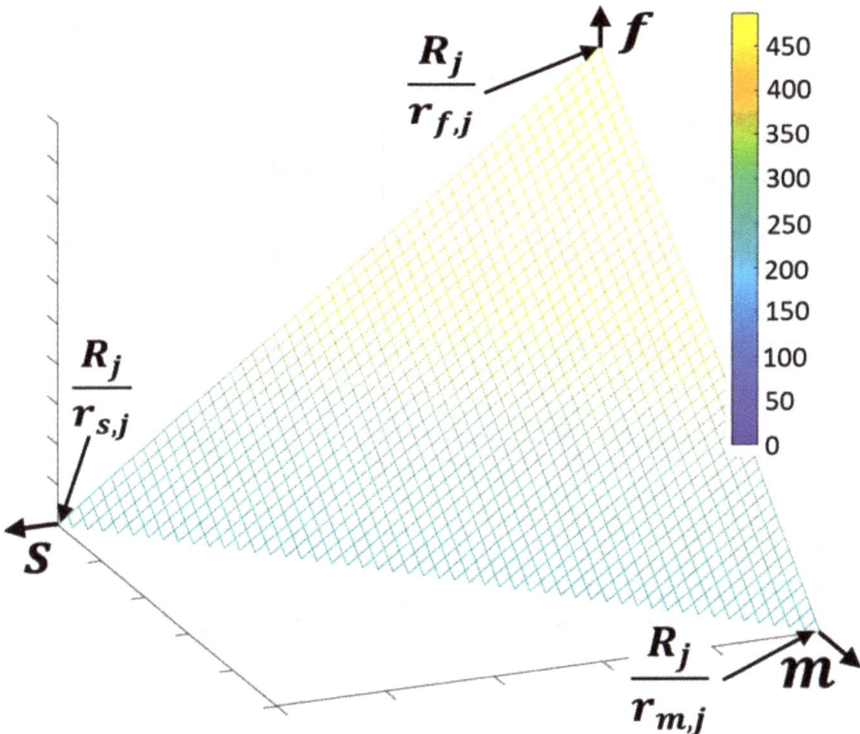

Figure 7. An example of the space of solutions as calculated by the model with highlighted boundaries.

After all the technically possible solutions are found, CS costs are applied, in order to optimize the system and find the least-cost solution for each energy level; this is done by searching the minimum value of the $\left(C_{TOT,j}\right)_n$ term of the equation:

$$\left(C_{TOT,j}\right)_n = C_s * s_{j,n} + C_m * m_{j,n} + C_f * f_{j,n}, \ \forall \left(s_j, m_j, f_j\right)_n. \tag{12}$$

The outputs of this step are the minimum cost of a charging infrastructure suitable for the j-th class charging request and its composition in terms of CS:

$$\left\{ \begin{array}{l} (C_{MIN})_j \\ (s_{MIN}, m_{MIN}, f_{MIN})_y \end{array} \right. , j \in [1, 10]. \tag{13}$$

2.2.5. IV Step: Output Definition

This final step is designed to aggregate the outputs coming from steps II and IV, and to post-process them with some of the inputs in order to obtain the other PI for the specific charging infrastructure. In this section the operations needed to accomplish the first goal will be described, while the results of the latter will be discussed in the next section.

BEV and PHEV daily charging on publicly available infrastructure:

The total number of xEV daily using the charging infrastructure during y-th year can be obtained by summation of the values related to every j-th level:

$$\begin{cases} (BEV_P)_y = \sum_{j=1}^{10} \left(BEV_j\right)_y \\ (PHEV_P)_y = \sum_{j=1}^{10} \left(PHEV_j\right)_y \end{cases} . \tag{14}$$

Energy and power demand from publicly available charging infrastructure:

Couples of values $\left(E_j, BEV_j\right)_y$ and $\left(E_j, PHEV_j\right)_y$ are sufficient for the calculation of total energy provided by charging infrastructure during y-th year:

$$(E_{TOT})_y = \sum_{j=1}^{10} \left(E_j * BEV_j\right)_y + \left(E_j * PHEV_j\right)_y. \tag{15}$$

In order to calculate the energy provided by each power level of the infrastructure, each of the j-th level contributions must be evaluated separately and finally summed:

$$[(E_{TOT})_k]_y = \sum_{j=1}^{10} k_{MIN,j} * E_j * r_{k,j}, k \in (s, m, f). \tag{16}$$

Publicly available charging infrastructure composition and cost:

The number of charging stations for the various CS power levels and their cost are provided as output for each y-th year and each j-th energy class by the equations state da the end of Section 2.2.4; total yearly values are obtained by a sum in j and total global values are the obtained by another sum in i:

$$\begin{cases} (C_{MIN}) = \Sigma_i \sum_{j=1}^{10} \left[(C_{MIN})_j\right]_y \\ (s_{MIN}, m_{MIN}, f_{MIN}) = \Sigma_i \sum_{j=1}^{10} \left[(s_{MIN}, m_{MIN}, f_{MIN})_j\right]_y \end{cases} . \tag{17}$$

Energy and power demand from private residential charging infrastructure

The basic assumption regarding the use of private residential charging infrastructure is that each vehicle is assumed to be used and charged every day. This simplifying assumption is related to the fact that, being the stall private and related to the vehicle, this one will be parked there at least once in the day, ready to be charged. The equation describing the total average daily energy request is:

$$[(E_r)_A]_y = \sum_V \left\{ \left[1 - (\%xEV_P)_y\right] * (xEV)_y * (avc_V)_y * avd_A \right\}. \tag{18}$$

Private residential charging infrastructure cost:

The least cost for the private residential charging infrastructure is obtained using the following equation:

$$[(C_r)_A]_y = \sum_{i=1}^{y} \sum_V \left\{ \left[1 - (\%xEV_P)_y\right] * (xEV)_y * (C_r)_y \right\}. \tag{19}$$

The least cost assumption derives from the fact that no overlap in the use of publicly accessible and private residential charging infrastructure is modeled, while a certain share of it is expected.

2.3. Scenario Definition

A total of 18 working scenarios have been defined; they are the result of a three-step combination of various assumptions, as described below.

The first step is related to the two different geographical areas analyzed—here Florence and Bruxelles—and affects the following input variables:

- Average daily travel avd_{FI} and avd_{BXL}.
- Publicly accessible stalls PP_{FI} and PP_{BXL}.
- Private residential parking PR_{FI} and PR_{BXL}.

The second step considers the three different xEV fleet forecast scenarios calculated for each geographical area, here defined as low, medium and high; it affects the following inputs variables:

- BEV circulating fleet on y-th year $(BEV_{FI})_y$ and $(BEV_{BXL})_y$.
- PHEV circulating fleet on y-th year $(PHEV_{FI})_y$ and $(PHEV_{BXL})_y$.

The final step is related to user behavior modeling and specifically to the charging events probability distribution over the estimated range of the vehicle p.

3. Results and Discussion

3.1. Input Values Definition for the Implemented Scenarios

In the following sections the specific input values used in the model for this study were described. BEV and PHEV fleets forecast over analysis timeframe:

An historical dataset was needed as a baseline of auxiliary information for the development of the transfer formulas for dataset downscaling; the main sources for data collection are shown in Table 2, with a reference to the area considered.

Table 2. Data sources for historical baseline characterization (FI = Florence, BXL = Bruxelles).

Source	EU-28	IT	BE	FI	BXL	Data
Centro Studi Continental [A] [35]	-	-	-	X	-	$\frac{EV_{FI}}{EV_{IT}}$
Comune di Firenze [36]	-	-	-	X	-	TOT
EAFO [26]	X	X	X	-	-	NR_{EV}
ECOSCORE [B] [37]	-	-	X	-	X	TOT, EV
ENEL e-mobility [14]	-	X	-	-	-	EV
Eurostat [19]	X	X	X	-	-	N_{TOT}, N_{EV}
IBSA [38]	-	-	X	-	X	TOT, EV
UNRAE [39]	-	X	-	-	-	NR_{TOT}, NR_{EV}

[A]: Author's elaboration on ACI data; [B]: owned by Vito.be.

Table 3 highlights the sources used to collect the forecast data for xEV fleets growth. Since only [13,26,39] reported separately the contribution of BEV and PHEV, an average value of the allocation suggested by those papers has been used to divide the other forecast data between BEV and PHEV contributions.

A total of 23 scenarios, from nine different studies was thus selected. Then these were compared to the maximum assumed turnover for xEV, as described in Section 2.2.1 and only the ones that were proposing forecasts lower than the maximum assumed turnover for xEV were used in the model. After this step, 11 scenarios from seven sources remained; Figure 8 shows the results in terms of xEV shares over the total circulating fleet, for Florence municipality. It also highlights the portion of scenarios range that overcame the selecting threshold.

Table 3. Data sources for xEV fleet size forecasts.

Source	Paper	Analyzed Area	Release Year	Proposed Scenarios			
				N.	2020	2025	2030
Eurelectric *	Smart Charging: steering the charge, driving the change [40]	EU-27	2015	3	X	X	X
Delft-CE	Impact analysis for market uptake scenarios and policy implications [41]	EU-27	2011	3	X	X	X
RSE	Roadmap per una mobilità sostenibile [42]	IT	2017	1	X	X	X
Start/CEI CIVES	Libro Bianco EV [43]	IT	2017	4	X	-	-
EAFO	The transition to a Zero Emission fleet for cars in the EU by 2050 [26]	EU-28	2016	3	X	X	X
ENEL/Ambrosetti	E-Mobility Revolution [14]	IT	2017	4	-	X	X
PoliMi	E-Mobility report 2018 [13]	IT	2018	3	X	X	X
ENTSO-E	TYNDP 2018 [44,45]	IT, BE	2018	3	X	X	X
European Commission	Clean Power for Transport [46]	IT, BE	2017	3	X	X	X

*: 2020 to 2030 data obtained by linear interpolation.

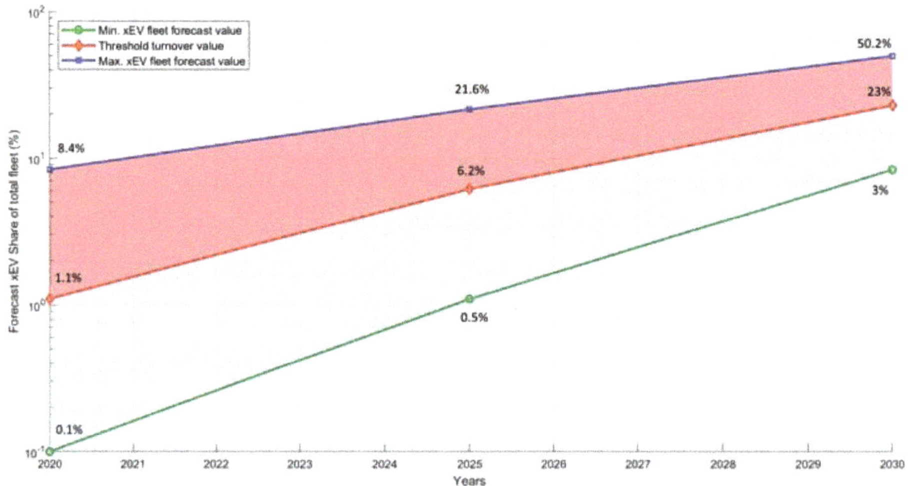

Figure 8. xEV fleet growth forecast—total range of collected data values and highlight of the selected part (light blue), under threshold limit (red line).

Finally, Table 4 highlights the forecasts for BEV and PHEV circulating fleets during the considered time period, used in the model as inputs for this research.

Table 4. Battery electric vehicle (BEV) and plug-in hybrid electric vehicle (PHEV) fleet forecasts for the area of Florence and Bruxelles municipalities—absolute figures.

Area	Scenario	BEV			PHEV		
		2020	2025	2030	2020	2025	2030
Florence	Low	73	441	2062	90	517	2544
	Medium	202	1414	6344	584	2980	9886
	High	672	3968	14,030	1141	6016	21,532
Bruxelles	Low	663	3482	10,334	984	3482	12,168
	Medium	1283	6022	18,432	2655	10,266	25,451
	High	2689	10,078	35,836	5152	17,591	45,337

Table 5 shows the same data but this time in terms of shares of the total circulating fleets in the two selected areas.

Table 5. BEV and PHEV fleet forecasts for the area of Florence and Bruxelles municipalities—shares of total circulating fleet.

Area	Scenario	BEV			PHEV		
		2020	2025	2030	2020	2025	2030
	Low	0.0%	0.2%	1.0%	0.0%	0.3%	1.3%
Florence	Medium	0.1%	0.7%	3.2%	0.3%	1.5%	5.0%
	High	0.3%	2.0%	7.0%	0.6%	3.0%	10.8%
	Low	0.1%	0.7%	2.1%	0.2%	0.7%	2.5%
Bruxelles	Medium	0.3%	1.2%	3.7%	0.5%	2.0%	5.2%
	High	0.5%	2.0%	7.3%	1.0%	3.4%	9.2%

Average BEV and PHEV energy consumption and capacity of batteries:

Table 6 shows the results obtained by applying the methodology described in Section 2.2.1 and subsequently applied in the model.

Table 6. xEV average consumption and capacity assumed forecast values.

	2020		2030	
	BEV	PHEV	BEV	PHEV
Average specific consumption (kWh km^{-1}) $(avc_V)_y$	0.192	0.108	0.154	0.086
Average battery useful capacity (kWh) $(batt_V)_y$	31.6	7.14	59.5	9.8

Public and private stalls availability:

In order to obtain the results highlighted in Table 7 the following sources were investigated:

- For the Florence area, open-data maps and databases developed by the municipality were used to assess public parking spaces availability [47], while the results of the 2011 census [48] were used to evaluate private residential parking spaces.
- With regards to the Bruxelles area, open-data databases were used to obtain the number of public parking spaces available, together with an estimate of private parking spaces in the residential area [49].

Table 7. Public and private stalls availability in the analyzed areas of Florence (2013 data) and Bruxelles (2018 data).

	Public Stalls (PP_A)	Private Residential Parkings (PR_A)
Florence	65,000	55,800
Bruxelles	318,600	293,000

Definition of CS characteristics:

As a first step in choosing the implemented CS power level a research within existing National Regulations and EU-28 Directives has been performed, in order to gather suggestions and any provided prescription. The main sources consulted here were Directive 2014/94/EU [4] and the PNire [5].

Another research has then been performed to understand the current composition of publicly accessible charging infrastructures in the study areas, using accessible databases such as official open-data archives of municipalities [50], data from a service provider (Enel) [51] and other unofficial archives (Opencharger) [52], to understand the current composition of the infrastructure in terms of quantity, power levels and connection standards used.

The home charging sector has also been considered, evaluating the average installed power as well as the maximum that can be installed by a European household, to define an upper limit value for the power of the recharging system [53]. In conclusion, a literature research was carried out in order to evaluate the possible future developments of the infrastructure in terms of power levels and usage [34].

At the end of the research phase, the following power values were identified for the various charging systems used in the model, and finally applied both to Florence and Bruxelles area:

- Private residential charging points P_r: 2 kW
- Publicly accessible charging points: Three power levels and corresponding charging systems, called slow, medium and fast, have been implemented. The selected power values are listed in the following Table 8:

Table 8. Charging power levels assumed in the model for this study.

Power Levels	Charging Power	
	2020–2024	2025–2030
Slow—$(P_s)_y$	7 kW	22 kW
Medium—$(P_m)_y$	22 kW	50 kW
Fast—$(P_f)_y$	50 kW	100 kW

An installation cost has been defined for each CS power level; this value includes the cost of the charging station, including installation and grid connection costs. Operating costs were not considered, as they were not part of the scope of the study, nor were maintenance costs; the latter were not implemented in order maintain the model simple enough.

Learning curves were applied to capital, installation and connection costs, to take into account the reduction of costs over time related to the effects of industrialization and rationalization of processes. All the information derives from an extensive literature research (cfr. [10,31,41,46,54,55]) and brought to the definition of $(C_s)_y$, $(C_m)_y$, $(C_f)_y$ and $(C_r)_y$, related to the y-th year and described in Table 9.

Table 9. Charging station (CS) costs forecast for the various power levels.

	2020	2025	2030
Private Residential—2kW $(C_s)_y$	900 €	810 €	729 €
Slow—7 kW $(C_m)_y$	1783 €	1630 €	1630 €
Medium—22 kW $(C_f)_y$	4800 €	4264 €	4264 €
Fast—50/100 kW $(C_r)_y$	44,000 €	38,798 €	32,564 €

A literature search was then carried out to verify the actual use ratios of existing charging infrastructures [20,56]. The driving and parking behaviors and daily timeframes of an average user were mainly derived from the study [23], which specifically analyses driving behavior in various European countries, including Italy. The results of these evaluations, together with other usage constraints and the energy request levels E_j, brought to the maximum number of charging events for level of energy requests reported in Table 10 for each power level of the CS:

Driving and charging behavior of users:

For what it concerns the average daily travelled distance, two different values have been calculated, for Florence and Bruxelles, as shown in Table 11 the most recent data available (2015) have been used, from:

- Odyssee-Mure Database [57]: Italy and EU-28.
- ACEA [58]: Belgium.
- ACI—CENSIS [59]: Italy.

Table 10. Maximum number of charging events daily manageable by each level of energy requests, for each Charging energy request class.

CS Power Levels		Charging Energy Request Classes—E_j (kWh)									
		6	12	18	24	30	36	42	48	54	60
2 kW	$r_{r,j}$	1	1	1	0	0	0	0	0	0	0
7 kW	$r_{s,j}$	3	3	3	2	2	2	2	2	2	0
22 kW	$r_{m,j}$	8	8	8	4	4	4	4	3	3	3
50 kW	$r_{f,j}$	12	12	12	12	8	8	8	8	4	4
100 kW	$r_{f,j}$	17	17	17	12	12	12	12	8	8	8

Table 11. Forecast average daily travelled distance by a passenger car over the analyzed areas.

	Florence	Bruxelles
Average daily travelled distance (km) (avd_A)	23.5	30.5

The shares of PHEV and BEV, which are considered as using the publicly accessible charging infrastructure, were calculated using data coming from [31,60]; the results are reported in Table 12:

Table 12. Share of PHEV and BEV expected to be using the publicly accessible charging infrastructure.

	2020	2025	2030
Share of BEV using the publicly accessible charging infrastructure ($\%BEV_P)_y$	56.5%	66.9%	75.8%
Share of PHEV using the publicly accessible charging infrastructure ($\%PHEV_P)_y$	45.6%	58.6%	69.7%

Finally, the charging events probability distribution p over the estimated range of the vehicle, was defined for three different scenarios (see Figure 3):

- Linear: assumes a linear correlation between the number of days since the last charging events and the charging event probability.
- Regressive: assumes a correlation between the number of days since the last charging events and the charging event probability of the kind $y = \sqrt{x}$.
- Progressive: assumes a correlation between the number of days since the last charging events and the charging event probability of the kind $y = x^{5/3}$.

3.2. Discussion

The analytical model described in this article was designed to capture the main complexities of the selected scenarios, while maintaining a simple and lean structure. Therefore, some of the simplifying assumptions implemented in the model could affect the outputs and the accuracy in some specific areas.

As already stated in Section 2.1, the model did not update the utilization ratio of the already deployed charging infrastructure to take into account incremental improvement during the time period. This could potentially lead to a sub-optimal use of the charging infrastructure. Overall, infrastructure oversizing can be estimated in just a few percentage points from optimum, since it is directly proportional to utilization ratio. On the other hand, the model evaluates the average daily values, thus simulating an average use of the charging infrastructure; this could lead to an underestimation of peak power demand from the grid, as well as to a possible underestimation of the number of CS needed by the circulating xEV fleet. Finally, the model has to deal with the lack of real data related to charging behaviors of users, and with the uncertainty of the forecasts used as inputs; this situation clearly leads to a corresponding range of variability of the outputs.

All the following outputs will be thus presented in a min–max format, given also the fact that they will summarize nine different analyzed scenarios for each geographical area. Moreover, it must be highlighted that the maximum values refer to a threshold situation related to the turnover trends of passenger vehicles circulating fleet and thus have to be considered as an "upper limit". That said, the main results related to public charging infrastructure are reported in Table 13, while the main results related to private charging infrastructure are highlighted in Table 14, both for Florence and Bruxelles area.

Table 13. Model results for xEV fleet forecast and publicly accessible charging infrastructure for Florence and Bruxelles area.

Output (Cumulative Values)		Publicly Accessible Charging Infrastructure					
		Florence			Bruxelles		
		2020	2025	2030	2020	2025	2030
EV fleet/Tot fleet (%)	Min	0.1%	0.5%	2.3%	0.3%	1.4%	4.6%
	Max	0.9%	5.0%	17.8%	1.5%	5.4%	16.5%
Total CS	Min	12	66	194	201	580	1210
	Max	199	795	2053	1014	2538	5537
Total costs (k€)	Min	24.4	169.6	743.7	369.2	1366.2	4496.5
	Max	385.1	2309.5	8323.9	1959.1	6993.4	21,381.9
Average daily charged energy (kWh)	Min	0.3	2.2	11.0	4.3	20.4	68.7
	Max	3538	21,317	81,363	19,514	74,215	244,712
Global utilization rate	Min	10.61%	9.59%	8.88%	11.70%	10.77%	10.27%
	Max	16.60%	14.77%	14.34%	14.69%	16.19%	14.44%
EV/CS ratio	Min	10	12	18	8	11	15
	Max	18	19	27	9	13	19
"Slow [A]" CS/"Fast" CS ratio	Min	n.a. [B]	59	188	n.a.	166	31
	Max	n.a.	n.a.	n.a.	n.a.	n.a.	72

[A] Here 7kW and 22kW power levels are assumed as "Slow"; [B] "n.a.", whenever present, means that the corresponding charging infrastructure does not have 50 kW and 100 kW CS.

The average amount of daily energy required by xEV fleet charged on public infrastructure is assumed to reach more than 80 MWh for Florence and almost 245 MWh for Bruxelles by 2030, for the most demanding scenario. These values correspond to an average daily electricity consumption of about 13,500 families in Florence, and about 35,000 families in Bruxelles [53]. Considering an average delivery window of 20 h, compatible with the assumptions made in the study, this energy demand theoretically corresponds to an average continuous power requirement of 4 MW for Florence and 12.3 MW for Bruxelles.

The same analysis, when applied to the private residential charging infrastructure, give as results for the 2030 values of almost 50 MWh for the Florence and almost 155 MWh for Bruxelles for the most demanding scenario. Considering an average delivery window of 8 h, compatible with the assumptions made in the study of only night charges, this energy requests corresponds to an average continuous power requirement of 6.3 MW for Florence and 19.4 MW for Bruxelles.

These outputs show that the share of total energy demands supplied by private residential charging infrastructure ranged between slightly more than 50% on year 2020 to less than 40% at the end of the time period, on year 2030. These results were strictly related with the assumption on the share of xEV using private infrastructure defined in Table 12.

Table 14. Model results for xEV fleet forecast and private residential charging infrastructure for Florence and Bruxelles area.

Output (Cumulative Values)		Florence			Bruxelles		
		2020	2025	2030	2020	2025	2030
Total CS	Min	85	442	1676	877	3285	8559
	Max	967	4694	13,401	4191	13,253	31,386
Total costs (k€)	Min	76.5	367.8	1267.4	1078.2	3736.4	8917.4
	Max	870.3	3910.2	10,257.6	3771.9	11,163.1	24,382.1
Average daily charged energy (kWh)	Min	365	1828	6318	4778	18,033	42,722
	Max	3962	18,739	49,524	22,047	67,663	154,527
Share of total charging energy	Min	52.1%	45.6%	36.5%	52.9%	46.9%	38.3%
	Max	52.8%	46.8%	37.8%	53.0%	47.7%	38.7%

The header row above the Florence/Bruxelles split reads "Private Residential Charging Infrastructure".

Another interesting perspective for the comparison of both private and publicly accessible infrastructure was the total cost of installation: private infrastructure results were always more expensive, up to twice the cost for some scenarios. This poor performance was related to the lowest EV/CS ratio of residential CS that was assumed to be equal to 1, so that basically for every xEV that uses the private infrastructure a CS is needed.

Finally, with the aim of verifying the effort to be made to achieve the results described by the model, the average number of charging stations to be installed each year and the related annual cost were then calculated. The results were averaged over the two periods 2020–2025 and 2025–2030 for the areas of Florence and Bruxelles and are shown in Table 15.

Table 15. Yearly steps to deploy the publicly accessible charging infrastructure as sized by the model.

Output (Annual Mean Values)		Florence		Bruxelles	
		2020–2025	2025–2030	2020–2025	2025–2030
CS to be installed in a year	Min	11	23	108	112
	Max	143	223	464	536
Yearly deployment costs	Min	25,387 €	102,789 €	238,265 €	550,855 €
	Max	373,816 €	1,075,794 €	1,192,045 €	2,570,283 €

After this brief overall presentation of the results, in the following sections more in-depth analyses would be presented on specific arguments.

Publicly accessible infrastructure composition:

The most important result obtained from the analysis of model outputs was that the higher the power CS was the smaller share of the total, as Figure 9 shows. This was related to their worst charging events/cost ratio $r_{k,j}/(C_k)_y$ comparing to slow and medium ones. In other words, the high-power CS was able to manage less daily charging events for the cost.

Moreover, an inversion trend between slow and medium ones was highlighted during the timeframe of the analysis. This behavior was related to the shift to higher E_j classes of the energy requests moving through the time period. The energy distribution pattern between the various power level also confirmed these findings, showing an even stronger predominance of the medium, 22 kW, CS power level.

Number of daily charging xEV as a percentage of the total EV fleet:

A consequence of vehicle range and user behavior assumptions, only a share of xEV ranging from 34% to 89% of theoretical total value $\left[(BEV_P)_y + (PHEV_P)_y\right]$ used the charging infrastructure on an average, daily basis, with the expected trend shown in Figure 10. This finding directly reflected into less charging events with average higher energy requests. This could lead to a smaller than expected infrastructure but with the same average energy request thus, consequentially, possible higher power request.

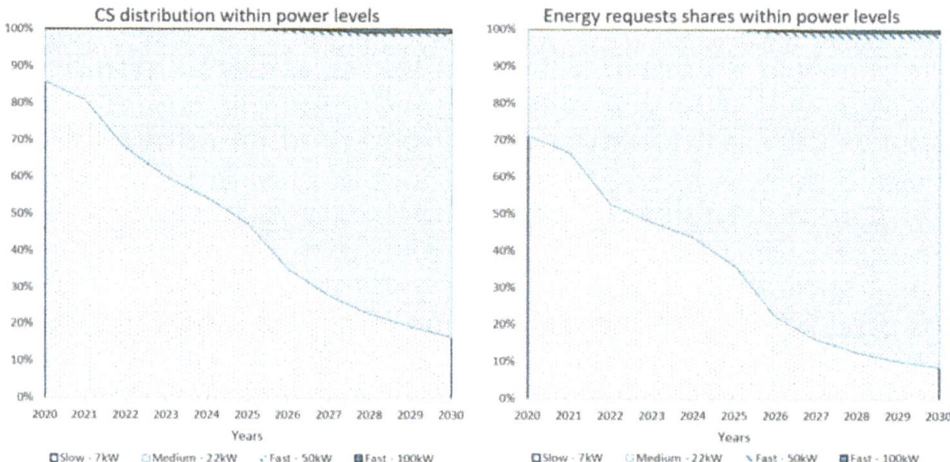

Figure 9. CS distribution (left) and energy requests distribution within power level (right) for the publicly accessible charging infrastructure.

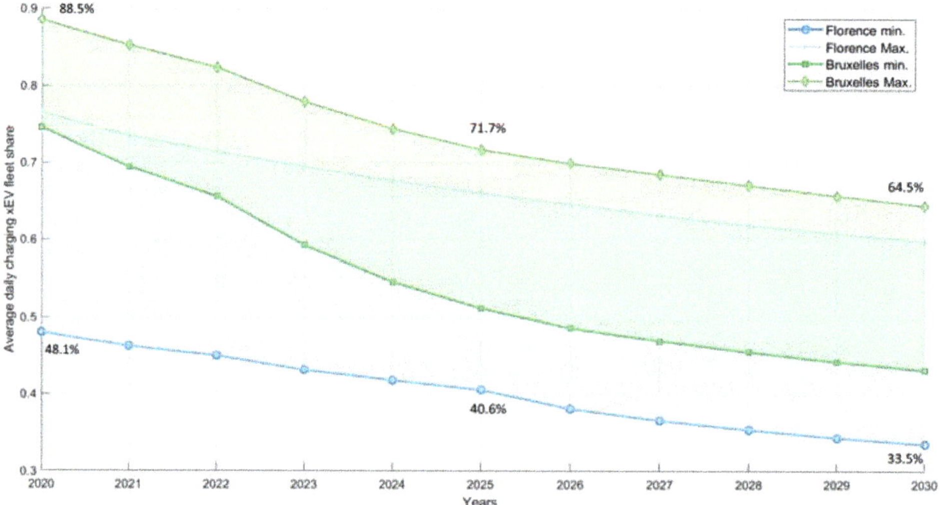

Figure 10. Publicly accessible charging infrastructure daily charging events for Florence (light blue) and Bruxelles (light green).

The higher values calculated for Bruxelles area could be related to the higher circulating fleet, thus to the higher number of PHEV; in fact, PHEV were modeled to a much smaller battery capacity, so they usually charge more often.

EV/CS ratio:

This parameter describes the number of xEV virtually assigned to each existing CS of the infrastructure. Higher EV/CS led to costs reduction for the infrastructure and higher revenues for each

CS; anyway, also negative side effects could occur, such as a higher risk of finding the CS occupied by other charging xEV. A balance between these aspects were thus found; AFI Directive suggests a 10:1 ratio [4], while Italian PNIRE suggest an average figure of 7:1 [5].

The following equation was used within the model to define this parameter:

$$(EV/CS)_y = \sum_{i=1}^{y} \left\{ \frac{(BEV_P)_i + (PHEV_P)_i}{\left[\sum_j (s_{MIN} + m_{MIN} + f_{MIN})_j\right]_i} \right\}. \tag{20}$$

A rising trend is shown in Figure 11, going through the timespan of the analysis; this is due to a growing path in the installation of higher power CS—such as medium and fast ones—capable of higher numbers of daily charging event. Figure 11 shows also that EV/CS ratio figures were spread over a broader range and were also higher on average for Florence, when compared to Bruxelles. This again was related to the higher number of PHEV expected to be circulating in Bruxelles, given the fact that their lower energy requests were optimally fulfilled by the CS of the lowest power level, which were also the one with the lowest EV/CS ratio. The higher variability of the xEV fleet size forecast for Florence probably affected also the variability of EV/CS ratio for this area.

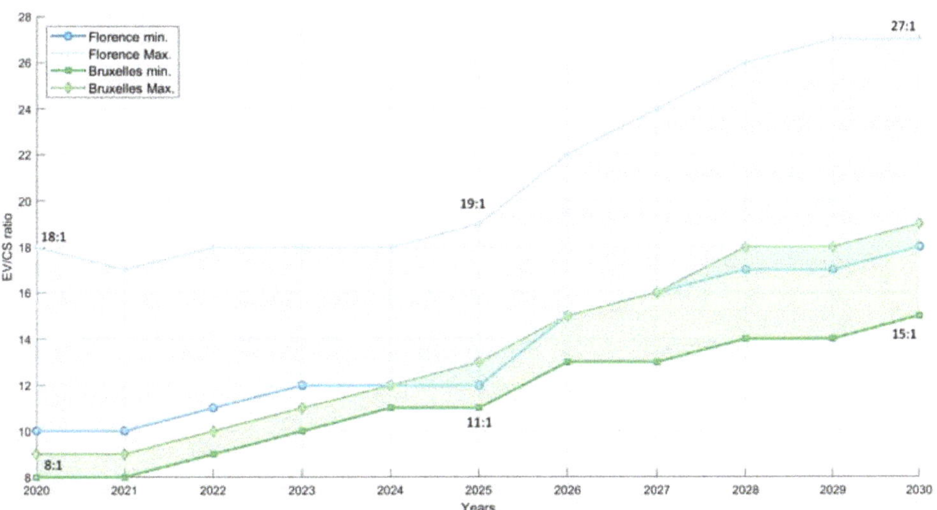

Figure 11. Publicly accessible charging infrastructure electric vehicle (EV)/CS ratio for Florence (light blue) and Bruxelles (light green).

Global use ratio of infrastructure:

An important parameter for business and profitability evaluations is the global usage ratio u, which is defined within this model using the following equation:

$$u_y = \frac{\sum_k [(E_{TOT})_k]_y}{\sum_k (k_{MIN})_y * h_k * (P_k)_y}. \tag{21}$$

Figure 12 shows that the global use ratio was quite low and rather constant during the whole period; the results were in line with other researches findings, such as [20,22,56].

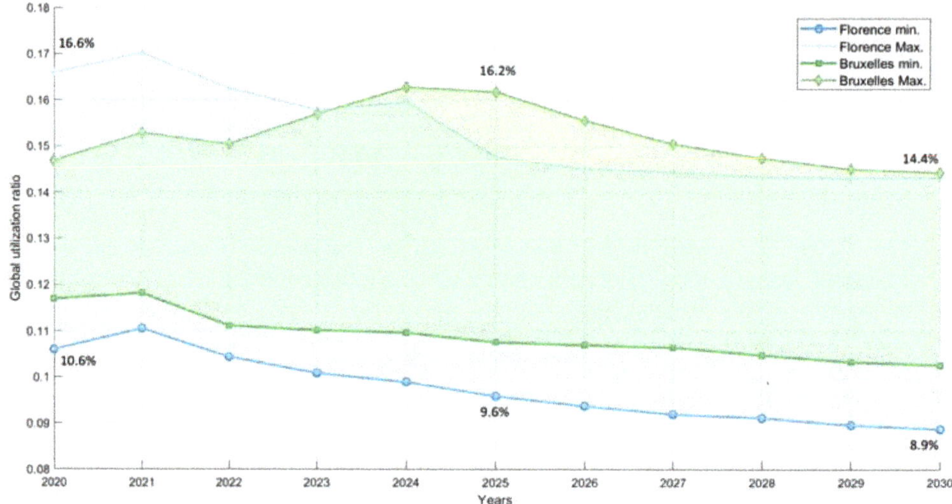

Figure 12. Publicly accessible charging infrastructure utilization rate for Florence (light blue) and Bruxelles (light green).

Impact of infrastructure on private and public parking:
The impact on public stalls I_y for a given area is defined as following:

$$I_y = \frac{CS_y}{\frac{xEV_y}{TOT_Y} * PP}.$$

(22)

The resulting value was low and the trend decreasing in time, as highlighted by Figure 13; this was the effect of the high value of EV/CS ratio, that also increased in the second part of the period given the higher share of medium and fast CS.

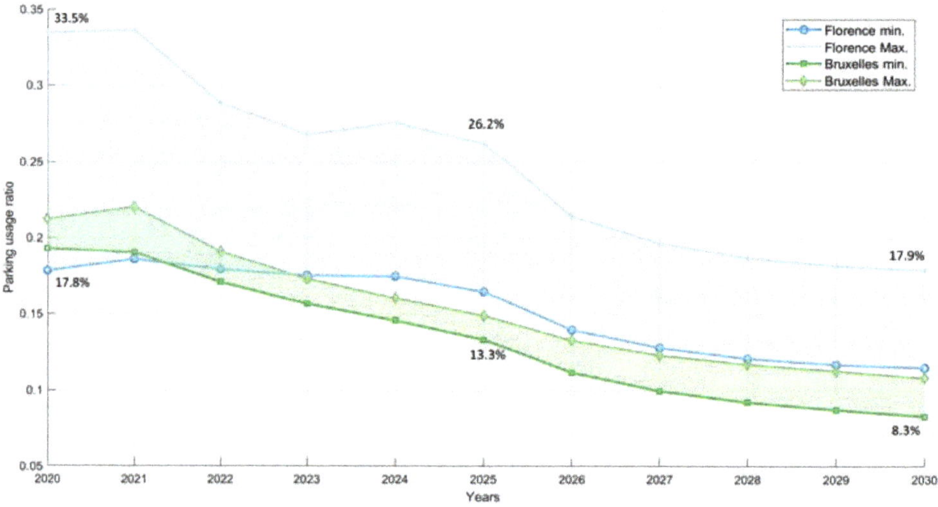

Figure 13. Public stall usage ratio by publicly accessible charging infrastructure for Florence (light blue) and Bruxelles (light green).

The impact on the private infrastructure was higher, given the fact that the probability of an xEV owner being also a parking owner was twice as high than with an ICEV owner; anyway, the result was still compatible with the existing situation.

As summarized in the following conclusions, the analysis carried out in this research work clearly identified elements impacting on policy at different governance levels. The central government layer, in charge of accelerating the transition to a cleaner transport system and implementing the EU Directives, will have to allocate the financial resources in the best possible way to allow local governments to maximize the fast deployment of charging infrastructure "on the ground". In this respect, the analysis here carried out about the optimal ratio between slow and medium CS to fast CS is a key element: results of our analysis did not seem to fully support the widely adopted idea (at the higher centralized policy level) of favoring fast CS versus slow and medium ones, in regards to cost optimization.

4. Conclusions

This work investigated a lean methodology to estimate the optimal size of a minimum cost charging infrastructure for passenger vehicles, suitable for deployment in an urban area. An analytical model was therefore developed to simulate different scenarios, accounting separately for BEV and PHEV fleets, as well as for publicly accessible and private residential CS. The model needs as inputs, for each timestep of the analysis, the number of daily charging events related to both BEV and PHEV fleets and the corresponding energy requests; moreover, it needs a full characterization of the CS in terms of expected performances and costs. A discrete-time Markov chain (DTMC), applied to a countable, finite state-space was used to obtain the estimate of the average daily distribution of the charging events within the circulating fleet, and the corresponding energy requests. Due to the primary importance of the inputs related to the size of the circulating EV fleet and since substantial differences between the various forecasts emerged from the research, a novel method for data sorting and conditioning of EV fleets forecasts was developed, using circulating fleet turnover rate as a threshold indicator.

The model was then applied to the two selected case studies of Florence Municipality for Italy and Bruxelles for Belgium, over the 2020–2030 period, with a 1-year timestep resolution; nine inputs scenarios and three outputs scenarios were defined for each area. The xEV fleet forecasts used as inputs for this work were presented with a broad range of variability: on 2030 they spanned between 2.3% and 17.8% of the total circulating fleet for the Florentine area, and between 4.6% and 16.5% for Bruxelles. In absolute terms, this translated to 194 to 2053 passenger cars for Florence and to 1210 to 5536 for Bruxelles. As a consequence of vehicle range and user behavior assumptions, only a share of xEV ranging from 34% to 89% of the theoretical total value was expected to use the charging infrastructure on an average day; this could lead to a smaller than expected infrastructure having the same average energy request thus, consequentially, a possibly higher power request.

The optimal size for the publicly accessible charging infrastructure, to be reached on 2030, was estimated in about 194 to 2053 CS for Florence and 1210 to 5537 CS for Bruxelles; these numbers corresponded to about 0.75 to 8.3 M€ on deployment cost for Florence and to 4.5 to 21.4 M€ for Bruxelles. On average, it was estimated that 10 to 200 CS has to be installed every year in Florence to comply with the deployment trend, with the yearly cost of deployment spanning between 25 k€ and 1 M€; these figures depended on the evaluated scenario and on the selected year of the time period. The same analysis, projected on Bruxelles, returned an estimate of 100 to 530 CS to be installed every year, for a cost ranging between 230 k€ and 2.6 M€ per year. The narrower range of values related to Bruxelles could be explained by the higher minimum level of xEV forecasted in comparison to Florence. Notably, the higher power CS results to be the smaller share of the total and this is clearly related to their worst charging events/cost ratio comparing to slow and medium CS. The private residential charging infrastructure size on 2030 was then estimated in 1700 to 13,400 CS and in 8600 to 31,400, respectively for Florence and Bruxelles, with deployment costs ranging from 1.3 M€ to 10.3 M€ and from 8.9 M€ and 24.4 M€.

These higher figures in terms of costs and number of CS are related to the lower EV/CS ratio of residential CS, that is assumed equal to be 1; moreover, these figures suggest the relevance of a possible implementation of support schemes for the installation of residential CS, since this could help to unlock the potential related to households with off-street private parking.

Within the publicly accessible infrastructure, EV/CS ratio results into higher values, spanning from 8:1 to 27:1 depending on scenarios and time period; these values were also higher than AFI Directive suggestions for a 10:1 ratio and Italian PNIRE suggestions for an average figure of a 7:1 ratio. Even with these high EV/CS ratios, the global use ratio of the infrastructure did not rise over 17% of its theoretical maximum potential, showing an almost constant trend over the period, thus highlighting the possible need for policies supporting CS profitability, at least during the first transition period. The impact of the planned infrastructure on the available public stalls results were less pronounced than that of the ICEV and showed a decreasing trend over the period, as a result of the increasing EV/CS ratio.

From a grid perspective, the average amount of daily energy required by xEV fleet charged on public infrastructure was expected to reach more than 80 MWh for Florence and almost 245 MWh for Bruxelles by 2030, for the most demanding scenario. As a comparison, the private residential charging infrastructure, was expected to reach 50 MWh for Florence and 155 MWh for Bruxelles; these results were strictly related with the assumptions on the share of xEV using private infrastructure, as defined in the model. Based on the assumptions made for the daily charging windows duration, energy requests translate into average continuous power requirements for the public infrastructure of 4 MW in Florence and 12.3 MW in Bruxelles; the same analysis made on the private residential charging infrastructure, gave as a result 6.3 MW for Florence and 19.4 MW for Bruxelles.

Finally, results shown in Section 3.2 made evident that optimal charging infrastructure configuration was obtained with a higher share—well beyond 90% of the total at 2020—of slow and medium CS, compared to fast ones. This situation was due to the worst charging events/cost ratio of the fast CS—if compared to the slow and medium ones—that is obtained from the input used in this study. It is worth noting that this result was quite different from the suggestion given by PNIRE of a 25% to 50% share of fast CS over the total on year 2020 [5].

Overall, therefore, more than the additional power demand by EV and the associated costs, the critical issue for developing a charging infrastructure able to meet the EV fleet on the coming years will most likely lie on the actual implementation of the civil works at urban level, and the ability to implement these vis-à-vis the EV fleet growth, so to achieve comparable development rates. This assuming a reasonable incremental rate of EVs in the next decades, thus excluding unrealistic and excessively optimist or pessimistic assumptions about car renewal rates.

Author Contributions: Conceptualization and Methodology, D.C.; Model Development, Methodology, Validation, Investigation, Formal Analysis, Writing and Editing: G.T.; Reviewing: D.C.; Supervision, D.C. and F.G.

Funding: This research received no external funding

Conflicts of Interest: Authors declare no conflict of interest.

Glossary

AFI	Alternative Fuels Infrastructure	GIS	Geographic Information System
ACI	Automobile Club d'Italia	ICEV	Internal Combustion Engine Vehicle
BEV	Battery Electric Vehicle	PHEV	Plug-in Hybrid Electric Vehicle
CS	Charging Station	PI	Performance Indicators
DTMC	Discrete Time Markov Chain	PNIRE	Piano Nazionale Infrastrutturale per la Ricarica dei veicoli alimentati ad energia Elettrica
EEA	European Environment Agency	RES	Renewable Energy Source
EU	European Union	RFNBO	Renewable Fuels of Non-Biological Origin
EV	Electric Vehicle	SOC	State Of Charge
GHG	Greenhouse Gases	xEV	BEV and PHEV

References

1. European Commission. *Statistical Pocketbook 2018. EU Transport in Figures. European Commission*; Publications Office of the European Union: Luxembourg, 2018; ISBN 9789279739514.
2. European Environmental Agency. *European Union Emission Inventory Report 1990–2016*; European Environmental Agency: København, Denmark, 2016; ISBN 9789292139506.
3. IEA. *World Energy Outlook 2016*; IEA: Paris, France, 2016; ISBN 978-92-64-26495-3.
4. The European Parliament. *DIRECTIVE 2014/94/EU of 22 October 2014 on the Deployment of Alternative Fuels Infrastructure*; European Parliament: Brussels, Belgium, 2014; p. 20.
5. Ministero delle Infrastrutture e dei Trasporti PNire. *Piano Nazionale Infrastrutturale per la Ricarica dei Veicoli Alimentati Ad Energia Elettrica*; Ministero delle Infrastrutture e dei Trasporti: Rome, Italy, 2016.
6. Energy-Transport Government Working Group. Belgian government national policy framework. *Altern. Fuels Infrastruct.* **2017**, *2*, 124.
7. Gkatzoflias, D.; Drossinos, Y.; Zubaryeva, A.; Zambelli, P.; Dilara, P.; Thiel, C. *Optimal Allocation of Electric Vehicle Charging Infrastructure in Cities and Regions*; JRC Science for Policy Report; European Union: Brussels, Belgium, 2016; p. 38.
8. Vanhaverbeke, L. *Location Modelling Approaches for the Optimal Roll-out of Charging Infrastructure*; Rotterdam, The Netherlands; VUB: Bruxelles, Belgium, 2016.
9. Cruz-Zambrano, M.; Corchero, C.; Igualada-Gonzalez, L.; Bernardo, V. Optimal location of fast charging stations in Barcelona: A flow-capturing approach. In Proceedings of the 10th International Conference on the European Energy Market (EEM 2013), Stockholm, Sweden, 27–31 May 2013.
10. Schroeder, A.; Traber, T. The economics of fast charging infrastructure for electric vehicles. *Energy Policy* **2012**, *43*, 136–144. [CrossRef]
11. Bhatti, A.R.; Salam, Z. A rule-based energy management scheme for uninterrupted electric vehicles charging at constant price using photovoltaic-grid system. *Renew. Energy* **2018**, *125*, 384–400. [CrossRef]
12. Nunes, P.; Figueiredo, R.; Brito, M.C. The use of parking lots to solar-charge electric vehicles. *Renew. Sustain. Energy Rev.* **2016**, *66*, 679–693. [CrossRef]
13. Energy & Strategy Group. *E-Mobility Report 2018*; Collana Quaderni AIP, Ed.; Energy & Strategy Group: Milan, Italy, 2018; ISBN 978-88-98399-28-4.
14. Ambrosetti S.p.A.; ENEL S.p.A. *E-Mobility Revolution*; ENEL: Rome, Italy, 2017.
15. Pruckner, M.; German, R.; Eckhoff, D. Spatial and temporal charging infrastructure planning using discrete event simulation. In Proceedings of the SIGSIM-PADS 17, Singapore, 24–26 May 2017; pp. 249–257.
16. De Gennaro, M.; Paffumi, E.; Martini, G. Customer-driven design of the recharge infrastructure and Vehicle-to-Grid in urban areas: A large-scale application for electric vehicles deployment. *Energy* **2015**, *82*, 294–311. [CrossRef]
17. Schuler, D.; Gabba, G.; Küng, L.; Peter, V. How a city prepares to e-mobility in terms of public charging infrastructure case study—The city of Zurich. In Proceedings of the 2013 World Electric Vehicle Symposium and Exhibition (EVS27), Barcelona, Spain, 17–20 November 2013; pp. 1–7.
18. Pagani, M.; Chokani, N.; Abhari, R.S.; Korosec, W. Techno-economic optimization of EV charging infrastructure incorporating customer behavior. In Proceedings of the International Conference on the European Energy Market (EEM 2018), Lodz, Poland, 27–29 June 2018; pp. 1–5.
19. Eurostat. Eurostat-Transport Database. Available online: https://ec.europa.eu/eurostat/web/transport/data/database (accessed on 3 January 2018).
20. Wirges, J.; Linder, S.; Kessler, A. Modelling the development of a regional charging infrastructure for electric vehicles in time and space. *Eur. J. Transp. Infrastruct. Res.* **2012**, *12*, 391–416.
21. Imperial College London. *Deliverable 4.3-B3-Grid Impact Studies of Electric Vehicles*; Imperial College London: London, UK, 2014.
22. Wirges, J. *Planning the Charging Infrastructure for Electric Vehicles in Cities and Regions*; KIT Scientific Publishing: Karlsruhe, Germany, 2016; ISBN 9783731505013.
23. Pasaoglu, G.; Fiorello, D.; Martino, A.; Scarcella, G.; Alemanno, A.; Zubaryeva, A.; Thiel, C. *Driving and Parking Patterns of European Car Drivers—A Mobility Survey*; Joint Research Centre—IET: Petten, The Netherlands, 2012; ISBN 9789279277382.

24. The European Parliament. *DIRECTIVE 2007/46/EC Establishing a Framework for the Approval of Motor Vehicles and Their Trailers, and of Systems, Components and Separate Technical Units Intended for Such Vehicles*; European Parliament: Brussels, Belgium, 2007; p. 160.

25. Eurostat. Eurostat Transport Data. Available online: https://ec.europa.eu/eurostat/web/transport/data/database (accessed on 10 August 2018).

26. Witkamp, B.; van Gijlswijk, R.; Bolech, M.; Coosemans, T.; Hooftman, N. *The Transition to a Zero Emission Vehicles Fleet for Cars in the EU by 2050*; EAFO: Bruxelles, Belgium, 2017.

27. EAFO EV Statistics. Available online: http://www.eafo.eu/vehicle-statistics/m1 (accessed on 10 October 2018).

28. Tsiakmakis, S.; Fontaras, G.; Cubito, C.; Pavlovic, J.; Anagnostopoulos, K.; Ciuffo, B. *From NEDC to WLTP: Effect on the Type-Approval CO_2 Emissions of Light-Duty Vehicles*; Publications Office of the European Union: Luxembourg, 2017; p. 50.

29. ACEA 2018 (Full Year) Europe: Electric and Plug-In Hybrid Car Sales per EU and EFTA Country-Car Sales Statistics. Available online: https://www.best-selling-cars.com/electric/2018-full-year-europe-electric-and-plug-in-hybrid-car-sales-per-eu-and-efta-country/ (accessed on 24 October 2019).

30. Boston, D.; Werthman, A. Plug-in Vehicle Behaviors: An analysis of charging and driving behavior of Ford plug-in electric vehicles in the real world. *World Electr. Veh. J.* **2016**, *8*, 926–935. [CrossRef]

31. Powell, N.; Hill, N.; Bates, J.; Bottrell, N.; Biedka, M.; White, B.; Pine, T.; Carter, S.; Patterson, J.; Yucel, S. *Impact Analysis of Mass EV Adoption and Low Carbon Intensity Fuels Scenarios*; Ricardo plc: Shoreham-by-Sea, UK, 2018.

32. Roscher, M.A.; Leidholdt, W.; Trepte, J. High efficiency energy management in BEV applications. *Int. J. Electr. Power Energy Syst.* **2012**, *37*, 126–130. [CrossRef]

33. Lutsey, N.; Meszler, D.; Isenstadt, A.; German, J.; Miller, J. *Efficiency Technology and Cost Assessment for U.S. 2025–2030 Light-Duty Vehicles*; ICCT White Paper; International Council on Clean Transportation: Washington, DC, USA, 2017; pp. 1–33.

34. IEA. *Global EV Outlook 2018*; IEA: Paris, France, 2018.

35. Centro Studi Continental. Centro Studi Continental Su Dati Aci. Available online: https://www.continental-pneumatici.it/auto/press/news/centro-studi/parco-circolante-auto (accessed on 10 October 2018).

36. Comune di Firenze. Available online: http://www.comuni-italiani.it/048/017/statistiche/veicoli.html (accessed on 10 August 2018).

37. Ecoscore. Ecoscore Belgium. Available online: http://ecoscore.be/en/fiches (accessed on 10 August 2018).

38. IBSA. IBSA-Transport. Available online: http://statistics.brussels/themes/mobility-and-transport#.W-VSS5NKguE (accessed on 24 October 2019).

39. UNRAE. *L'Auto 2018-Sintesi Statistica*; UNRAE: Rome, Italy, 2019.

40. Eurelectric. *Smart Charging: Steering the Charge, Driving the Change*; Eurelectric: Bruxelles, Belgium, 2015.

41. Grünig, M.; Witte, M.; Boteker, B.; Kantamaneni, R.; Gabel, E.; Bennink, D.; van Essen, H.; Kampman, B. *Impacts of Electric Vehicles—Deliverable 3 Assessment of the Future Electricity Sector*; CE Delft: Delft, The Netherlands, 2011.

42. RSE; MATTM; MIT. *Elementi Per Una Roadmap Della Mobilità Sostenibile Inquadramento Generale e Focus Sul Trasporto Stradale*; Editrice Alkes: Milan, Italy, 2017.

43. Guerriero, M.; Caretto, G.; Giardinelli, V. *Libro Bianco Sull'auto Elettrica*; START Magazine: Rome, Italy, 2017.

44. ENTSO-E. ENTSO-E-Maps. Available online: https://tyndp.entsoe.eu/maps-data/ (accessed on 11 September 2018).

45. Entso-E. *TYNDP 2018 Executive Report-Appendix*; ENTSO-E AISBL: Bruxelles, Belgium, 2018.

46. Wainwright, S.; Peters, J. *Clean Power for Transport Infrastructure Deployment—Final Report*; Publications Office of the European Union: Luxembourg, 2017.

47. Comune di Firenze-Stalli di Sosta. Available online: http://www.datiopen.it/it/opendata/Comune_di_Firenze_Stalli_di_sosta (accessed on 13 September 2018).

48. ISTAT. ISTAT-Censimento 2011. Available online: http://dati-censimentopopolazione.istat.it/Index.aspx# (accessed on 11 September 2018).

49. Open Data Brussels. Open Data Brussels-Stalls. Available online: http://opendatastore.brussels/ (accessed on 14 September 2018).

50. Open Data Brussels. Open Data Brussels-CS. Available online: http://opendatastore.brussels/dataset/charging-stations/resource/3355f77e-eb3d-45de-897b-5b2b5d10c4e5 (accessed on 11 October 2018).

51. ENEL S.p.A. Enel Drive. Available online: https://www.eneldrive.it/ (accessed on 10 October 2018).
52. Opencharger Opencharger Maps. Available online: https://map.openchargemap.io/ (accessed on 10 October 2018).
53. Odyssee-Mure. Odyssee-Mure Web Database-Indicators. Available online: http://www.indicators.odyssee-mure.eu/online-indicators.html (accessed on 13 November 2018).
54. Hall, D.; Lutsey, N.P. *Emerging Best Practices for Electric Vehicle Charging Infrastructure*; International Council on Clean Transportation: Washington DC, USA, 2017.
55. CREARA. *EUE 67–Cost of EV Charging Infrastructure*; CREARA: Madrid, Spain, 2017.
56. Funke, S.Á.; Gnann, T.; Plötz, P.; Funke, S.Á.; Gnann, T.; Plötz, Á.P.; Plötz, P. Addressing the different needs for charging infrastructure: An analysis of some criteria for charging infrastructure set-up. In *E-Mobility in Europe*; Springer: Berlin, Germany, 2015.
57. Odyssee-Mure. Odyssee-Mure-Travel Distances. Available online: http://www.odyssee-mure.eu/publications/efficiency-by-sector/transport/distance-travelled-by-car.html (accessed on 11 November 2018).
58. ACEA. *ACEA Report: Vehicles in Use Europe 2018*; ACEA: Bruxelles, Belgium, 2019.
59. ACI; CENSIS. *XX Rapporto ACI-CENSIS*; CENSIS: Rome, Italy, 2012.
60. CleanTechnica. Electric Car Drivers: Desires, Demands & Who They Are. 2016. Available online: https://cleantechnica.com/files/2017/05/Electric-Car-Drivers-Report-Surveys-CleanTechnica-Free-Report.pdf (accessed on 18 November 2018).

Article

PTG-HEFA Hybrid Refinery as Example of a SynBioPTx Concept—Results of a Feasibility Analysis

Franziska Müller-Langer [1,*], Katja Oehmichen [1], Sebastian Dietrich [1], Konstantin M. Zech [2], Matthias Reichmuth [3] and Werner Weindorf [4]

[1] DBFZ Deutsches Biomasseforschungszentrum gemeinnützige GmbH, Torgauer Straße 116, 04347 Leipzig, Germany; katja.oehmichen@dbfz.de (K.O.); Sebastian.dietrich@dbfz.de (S.D.)
[2] Deloitte GmbH, Rosenheimer Platz 4, 81669 München, Germany; Konstantin.zech@gmx.de
[3] IE Leipziger Institut für Energie GmbH, Lessingstraße 2, 04109 Leipzig, Germany; Matthias.reichmuth@ie-leipzig.com
[4] Ludwig-Bölkow-Systemtechnik GmbH, Daimlerstraße 15, 85521 München-Ottobrunn, Germany; Werner.weindorf@lbst.de
* Correspondence: franziska.mueller-langer@dbfz.de

Received: 15 August 2019; Accepted: 19 September 2019; Published: 27 September 2019

Featured Application: This work features a feasibility study of different operating scenarios for a hybrid refinery producing synthetic paraffinic kerosene from vegetable oil and electrolytic hydrogen.

Abstract: Limited alternative fuels for a CO_2-neutral aviation sector have already been ASTM certified; synthetic paraffinic kerosene from hydrotreated esters and fatty acids (HEFA-SPK) is one of these—a sustainable aviation fuel. With the hypothesis to improve the greenhouse gas (GHG) balance of a HEFA plant by realizing the required hydrogen supply via electrolysis—power to gas (PTG)—an exemplary SynBioPTx approach is investigated in a comprehensive feasibility study, which is, regarding this comparatively new approach, a novelty in its extent. About 10 scenarios are analysed by technical, environmental, and economic aspects. Within the alternative scenarios on feedstocks, electricity supply, necessary hydrogen supply, and different main products are analysed. For different plant designs of the hybrid refinery, mass and energy balances are elaborated, along with the results of the technical assessment. As a result of this environmental assessment, the attainment of at least 50% GHG mitigation might be possible. GHG highly depends on the renewability grade of the hydrogen provision as well as on the used feedstock. One important conclusion of this economic assessment is that total fuel production costs of 1295 to 1800 EUR t^{-1} are much higher than current market prices for jet fuel. The scenario in which hydrogen is produced by steam reforming of internally produced naphtha proves to be the best combination of highly reduced GHG emissions and low HEFA-SPK production costs.

Keywords: hybrid refinery; power-to-gas; biofuel; jet fuel; feasibility study

---

1. Introduction

The aviation sector is faced with particular challenges in regard to climate protection targets in light of the Paris Agreement, and at the same time it is faced with the rapid growth of the industry. Due to comparably long development and implementation phases in this sector as well as the long lifetime of aircrafts, the implementation of new powertrains (e.g., including batteries or fuels like hydrogen combined with fuel cells) usually requires several decades. Accordingly, sustainable aviation fuels (SAF) play a key role in the aviation sector. This is especially true for the so-called renewable drop in fuels that can be used similarly to conventional fossil jet fuel. Related to climate protection strategies, the demand on such renewable SAF is increasing in the medium-term [1,2]. Figure 1 shows the jet fuel demand of all flights departing in Germany, the CO_2 emissions to be reduced, and the share of required renewable jet fuels to fulfill targets of the Aviation Initiative for Renewable Energy in Germany e.V. (aireg), International Air Transport Association (IATA), and Climate Action Plan 2050 for Germany (CAP), as well as the required energy-related amounts of renewable SAF with specific greenhouse gas (GHG) mitigation potentials. According to this, a massive demand on SAF with high specific GHG mitigation compared to fossil jet fuel is required.

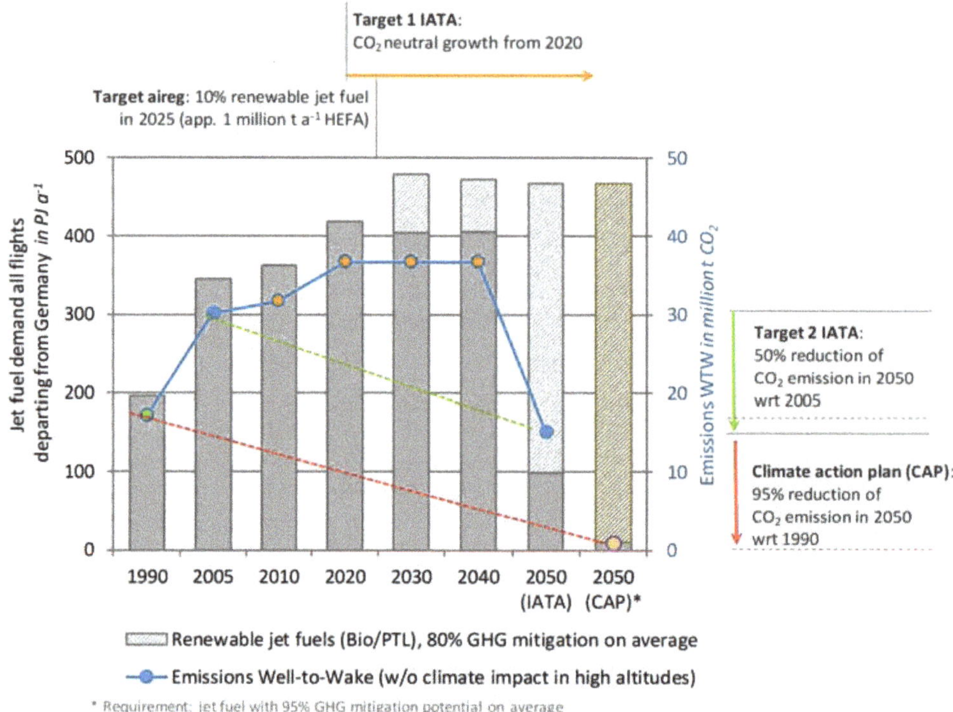

Figure 1. Jet fuel demand of all flights departing in Germany, emissions to be reduced, and the share of required renewable jet fuels to fulfill targets like of aireg, IATA, and climate protection plan (adapted from [3]).

Referring to the ASTM standards, which are binding in the aviation sector, only HEFA-SPK will short- to medium-term be available on the market in significant amounts [1]. Worldwide production capacities of HVO/HEFA (hydrotreated vegetable oils/esters and fatty acids) were about 224 Petajoule (PJ) in 2018 [4], of which about 5PJ was HEFA-SPK. According to ASTM D7566 Annex 2, HEFA-SPK can be blended to a maximum of 50% (v/v) to conventional jetfuel, which has been demonstrated several times but is not part of regular operation yet.

To produce HEFA, hydrogen is required. It was a hypothesis that renewable-based hydrogen might have a favorable GHG performance and—based on electrolysis (so called power-to-gas (PTG))—allow one to integrate renewable electricity. Therefore, it was the target when investigating a PTG-HEFA hybrid refinery as an example of SynBioPTx concepts in a comprehensive feasibility study. SynBioPTx stands for using synergies (syn) of biomass-based (bio) and electricity-based (PTx) fuels and product processes. This novel approach has got increasing interest in recent years, especially from the viewpoint of Germany where PTx fuels are seen as an important solution for climate-friendly transport and especially aviation (e.g. [5–9]).

Part of this first-of-its-kind feasibility study was technical, economic, and environmental assessments within a set frame given by the initiator of this study, like the size of the PTG-HEFA production facility to be realized in the time horizon 2024/25 and basic feedstock (here about 500,000 tonnes of jatropha oil). Considering Germany as a focus region for such a PTG-HEFA hybrid refinery (e.g., like using former mineral oil refinery sites), other sites in potentially favorable regions for PTG have also been investigated. Comparable studies that consider the combination of HEFA and PTG to this extent and in such detail are not known by the authors. Nevertheless, there are investigations on integrating renewable hydrogen into BTL routes (synthetic biomass-to-liquid) like [10,11].

The materials and methods are described, and results are shown and discussed as follows. Finally, conclusions are drawn.

2. Materials and Methods

The PTG-HEFA hybrid refinery has been assessed for different scenarios with regard to the plant setup. Technical assessment was done based on dedicated mass and energy balances, but also including issues on required infrastructure and different frame conditions. These were the base for the economic assessment with regard to costs and the environmental assessment with regard to GHG emissions.

2.1. Scenarios for PTG-HEFA Plant Setup

The technical feasibility of a PTG-HEFA hybrid refinery with a feedstock input of 500,000 tonnes per year of jatropha oil to HEFA-SPK was set as reference case to be realized in 2024/25. In addition to that, nine alternative scenarios have been investigated with regard to different approaches for electricity supply, hydrogen production, feedstocks, and refinery products. A simplified scheme of the PTG-HEFA hybrid refinery and the different frame options shows the scope (Figure 2).

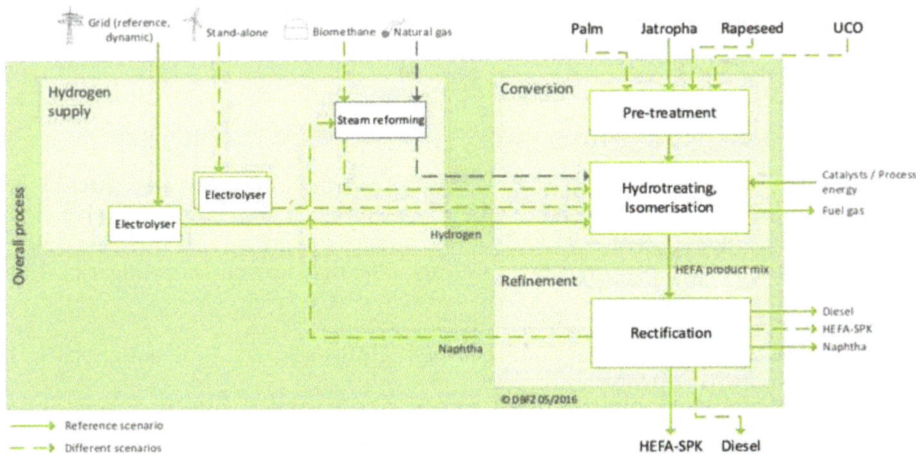

Figure 2. Simplified scheme of PTG-HEFA hybrid refinery and scope.

The specific assumptions for the different scenarios are summarized in Table 1. The different scenarios are grouped with special regard to the following:

(a) Electricity supply: constantly from the electricity grid with the specific electricity mix based on different primary energy sources, in a dynamic mode with based best cost-effectivity (in case of cheap spot market prices) whilst using electricity from the grid or as stand-alone power system based on fluctuating renewable energies (here, onshore wind park and solar photovoltaic park);

(b) Hydrogen supply: based on alkaline electrolysers with different storage systems depending on the electricity supply and—more conventionally—based on steam reforming from either natural gas, biomethane, or internal use of by-products of the HEFA plant such as naphtha and fuel gases (as is done today in most of the HVO/HEFA plants).

(c)   Feedstocks: as alternatives to jatropha oil; rapeseed oil with regard to the domestic option of, palm oil with regard to cost and specific hydrogen demand, and used cooking oils (UCO) with regard to lower GHG emissions (according the current regulations in the EU) .

(d)   Main product: referring to the different operational modes of a HEFA plant with regard to products, compared to the HEFA-SPK case, the alternative operation mode is focused on producing diesel.

Table 1. Scenarios and their specific assumptions.

Scenario	Feedstock	Electricity Supply	Necessary Hydrogen Supply (Electricity Demand in MW) [1]	Main Product
1 (reference)	jatropha oil	Constant from grid (with electricity mix): 8000 h a^{-1}	121 MW alkaline electrolyser incl. tube buffer storage	HEFA-SPK
2 (dynamic)		Dynamic from grid (spot market prices): 4000 h a^{-1}	244 MW alkaline electrolyser incl. salt cavern storage	
3 (stand-alone)		Stand-alone system (wind 360 MW and solar 250 MW): 2600 h a^{-1}	373 MW alkaline electrolyser with operated in varying part loads incl. salt cavern storage	
4 (natural gas)		Constant from grid (with electricity mix): 8000 h a^{-1}	via steam reforming from natural gas	
5 (biomethane)			via steam reforming from biomethane	
6 (naphtha)			via steam reforming from naphtha and fuel gas	
7 (rapeseed)	rapeseed oil		121 MW alkaline electrolyser incl. tube buffer storage	
8 (palm)	palm oil			
9 (UCO)	used cooking oil			
10 (diesel)	jatropha oil			diesel

[1] cf. calculations in Section 2.2.1.

2.2. Technical Assessment

As part of the technical feasibility, the plant conception of the PTG-HEFA hybrid refinery has been elaborated with regard to the different scenarios (Table 1) and—with regard to the assumed startup of such a plant in 2024/25—based on technologies that are and might be available on commercial scale. For all the scenarios, relevant mass and energy flows into and out of the plant have been calculated. Moreover, infrastructural issues with regard to possible locations for such a refinery have been investigated.

As a detailed description of the methods applied for the technical assessment of the PTG part and the HEFA part of the hybrid refinery is presented in [12]. Both parts are described briefly below.

2.2.1. PTG part

The power-to-gas part for the hydrogen supply is realized as alkaline electrolysis of water. The decision to include alkaline electrolysers into the hybrid concept was based on the results of a comprehensive market assessment (including concrete requests for proposal from international electrolyser manufacturers) that shows that only this technology can be provided by 2024/25 in the required capacity (Table 1). The different sizes are based on the calculations made for the different electricity supply scenarios that influence the annual operating hours of the electrolysers. The shorter the equivalent full load period of the electrolyser, the larger the capacity of the electrolyser to cover the hydrogen demand of the HEFA part (121 to 373 MW hydrogen including electricity for hydrogen compression). The hydrogen demand of the HEFA plant amounts to about 17,500 t a^{-1}. At an equivalent full load period of 8000 h a^{-1}, the electrolysis plant has to generate about 2.19 t h^{-1}, leading to an electricity input of 121 MW. At an equivalent full load period of 2600 h a^{-1}, the electrolysis plant has to generate about 6.66 t of hydrogen per hour, leading to an electricity input of 373 MW.

The same is true for the required type and size of the hydrogen storage (short-term underground tube buffer with a volume of 6860 m^3 with a net hydrogen storage capacity of 26 t for scenario 1 or salt caverns with a volume of 104,000/213,000 m^3 with a net hydrogen storage capacity of 775/1585 t for scenario 2 and 3), and thus the additional electricity demand for storage loading and unloading. The underground tube buffer is based on data from [13]. It has been assumed that for an equivalent full load period of 8000 h a^{-1} of the electrolysis plant, the required storage capacity amounts to about 0.5 days of full load operation leading to a net hydrogen storage requirement of about 26 t, which is approximately the net storage capacity of the underground tube gas storage described in [13]. Salt caverns only have been applied for scenario 2 and 3, where long-term hydrogen storage is required. Based on data in [14] and the equivalent full load period of the electrolysis plant, the required storage loading compressor capacity (about 2.19 t$_{H2}$ h^{-1} for an equivalent full load period of 4000 h yr^{-1} and about 4.47 t$_{H2}$ h^{-1} for an equivalent full load period of 2600 h yr^{-1}) and the required storage volume have been calculated. The hydrogen storage capacity amounts to about 15 days of full load storage loading compressor capacity.

2.2.2. HEFA part

The HEFA as multi-process plant (Figure 2) is derived from [15] and includes a pre-treatment of the feedstock (degumming, bleaching, and neutralisation), two main processing steps (hydrotreating, subsequent hydrocracking, and isomerisation stages), and a product separation (distillation).

All chemical reactions in the HEFA part (especially conversion of triglycerides of the feedstocks into linear, oxygen-free paraffins; saturation of double bonds; and hydrogenation on different paths and cracking) have been analysed on a molecular level according to the exact composition of each feedstock (data from [16–18]). This has been done in order to calculate the stoichiometric hydrogen demand per ton of feedstock according to [19,20], and the exact composition of the different fuel products.

The ratio of hydrodeoxygenation and decarboxylation in the conversion of the triglycerides is set to 73%:27% [21,22]. A calculated 30.2% of the CO$_2$ from decarboxylation is subsequently converted into methane by either direct conversion [23] or water–gas–shift reaction and CO methanation [24]. For each feedstock, the cracking rate is set individually depending on the requested product (HEFA-SPK or diesel).

The distribution of fuel products after the cracking stage is based on a normal distribution with a mean of half the initial chain length. The standard deviation is set according to the product distribution described in [25]. The process parameters for both process stages are set in reference to [26], and multiple cracking of one fatty acid is not taken into account.

As a state-of-the-art technology within HEFA facilities, in scenarios 4-6 the hydrogen production via steam reforming is considered instead of the PTG part. In the case of steam reforming, the total electricity demand of the plant is considerably reduced.

2.3. Environmental Assessment

The calculation of the lifecycle GHG emissions for the different scenarios has been conducted by means of a life cycle assessment (LCA), which is standardized and generally defined within DIN ISO 14040 and 14044 standards [27,28]. This general LCA approach described within these standards contains various levels of freedom regarding aspects such as system boundaries, life cycle impact categories, characterization factors, etc., which allow for a dedicated assessment. From a scientific point of view, these degrees of freedom are one of the strengths of LCA, but the results are difficult to compare. Proof of sustainability or, more specifically, evidence of a defined GHG reduction in a certification system requires a simplified method that allows for robust and consistent GHG accounting. For this purpose Annex V of the European Renewable Energy Directive (RED) contains an easy method for the calculation of lifecycle GHG emissions [29]. This method is based on the DIN EN ISO standards but limits the mentioned degrees of freedom by a clear definition of the system boundaries, the consideration of by-products and other aspects, which are described in Table 2.

Table 2. Methodology and assumptions for life cycle assessment (LCA).

Methodology Step	Assumptions for PTG–HEFA Hybrid Refinery
Goal and scope definition	
Considered impact categories	Global warming potential (GWP)
Functional unit	1 MJ fuel ex hybrid refinery
System boundary for LCA	Well-to-tank-chain including feedstock production (w/o direct or indirect land use changes), biomass collection of UCO and fuel production. No consideration of infrastructure (i.e., built up of plants, components, and vehicles)
Consideration of by-products	According to European Renewable Energy Directive (2008/29/EC) allocation of by-products (here, the subdivision of emissions and demands along the production chain between HEFA-SPK and naphtha, fuel gas, and diesel) and according to their energy content (lower heating value)
Inventory calculation	
Input/output analysis	Consideration of all relevant input and output streams (i.e., energy and feedstock inputs, auxiliaries and utilities, products and by-products, and waste) within the system boundary. Concepts based on process simulation, own data, and EcoInvent Version 3.3 [30]. External electrical power based on country specific mixes for 2015, emissions according to Gemis [31]
Impact assessment	
Approach	Evaluation of data resulting from input/output analysis regarding potential environmental impacts by means of so called characterising factor aggregation with regard to one reference substance
GHG emissions	According Forth IPCC Assessment Report (AR4) CH_4 with 25 CO_2-eq, N_2O with 298 CO_2-eq (w/o consideration of process-related biogenous CO_2 emissions) [32]
Result interpretation	
	cf. Section 3.2

2.4. Economic Assessment

As part of economic feasibility, specific fuel production costs have been calculated referring to the guideline 6025 of the Association of German Engineers' (VDI) [33] and by using the dynamic annuity model that takes all relevant cost items into account:

(a) Capital-linked costs: single and total investments for the different plant designs for the PTG-HEFA hybrid refinery;
(b) Consumption-linked costs: feedstocks, electricity, and auxiliaries (e.g., water, catalysts, natural gas and biomethane for steam reforming);
(c) Operation-linked costs: (plant staff, maintenance);
(d) Other costs: administration, insurance.

Revenues from by-products (here naphtha, fuel gas and diesel) are subtracted from the total costs. These total net-costs divided by the total amount of HEFA-SPK produced (or for scenario 10 for diesel) yield the specific fuel production costs.

A detailed method description for the economic assessment is presented in [12]. The main assumptions can be obtained from Table 3.

Table 3. Main assumptions for fuel production cost calculation (reference year 2015, assessment period 30 years).

Cost Factor	Value/Assumption	Reference
Total investment		
HEFA process units/incl. steam reformer	132/190 million EUR (annual load 8000 h)	[34–37]
electrolyser	58 million EUR (annual load 8000 h)	[38], quotations and
	116 million EUR (annual load 4000 h)	interviews
	176 million EUR (annual load 2600 h)	
hydrogen compression and storage	24 million EUR (annual load 8000 h)	[13,39]
	38 million EUR (annual load 4000 h)	
	70 million EUR (annual load 2600 h)	
Costs		
weighted average cost of capital	8% per year	[40]
maintenance HEFA part	2.5% of total investment per year	[34,35]
maintenance PTG part	9% of total investment per year	[38], quotations and interviews
administration, insurance, other	2.5% of total investment per year	[40]
personnel staff	50,000 EUR per year and person with full-time equivalent; app. 50 full-time equivalents required	[35,40,41]
feedstock (jatropha, rapeseed, palm, and UCO)	650/720/547/600 EUR t^{-1}	[4,42]
electricity (reference, dynamic, and stand-alone)	100/80/80 EUR MWh^{-1}	[12]
auxiliaries steam reformer (natural gas, biomethane)	480/896 EUR t^{-1}	[43] (biomethane)
..other auxiliaries (water, potassium hydroxide)	2/820 EUR t^{-1}	[40]
Revenues		
..naphtha, fuel gas	380/400 EUR t^{-1}	[44]
..diesel, jet fuel	410/425 EUR t^{-1}	[44]

2.5. Excursion on Favorable Regions

In addition to the framework analysed for a PTG-HEFA hybrid refinery in Germany, some other favorable regions with regard to high potentials for renewable electricity production have been investigated. Examples for that are Sweden, Spain and Namibia. For these regions, the scenarios 1 and 3 have been analysed with regard to GHG emissions and costs with a special focus on the PTG part of the hybrid refinery. For the stand-alone case with electricity supply from renewables wind power, photovoltaics (pv) and concentrated solar power (csp) have been considered. Details about the specific frame conditions there and the assumptions used for this feasibility analysis can be obtained from [3].

3. Results and Discussion

3.1. Technical Assessment

The results of the mass and energy balances are summarized in Table 4 for the different scenarios. Detailed discussion of these balances can be obtained from [12]. When jatropha is used as feedstock, the specific hydrogen consumption is about 35.7 kg per ton feedstock. For alternative feedstocks, this value is 39.9 kg t^{-1} for rapeseed; 31.6 kg t^{-1} for palm; 39.6 kg t^{-1} for UCO; and, in case of diesel as the main product, about 29.8 kg t^{-1}, with direct impact on the electricity demand for the PTG part (Table 4). The annual demand of potassium hydroxide for the electrolyser is about 940 kg. In case of steam reforming with external supplied natural gas or biomethane, about 67,200 tons per year are required; naphtha and fuel gas for steam reforming are provided plant-internally.

Table 4. Most important mass and energy balances for the different scenarios.

Scenario	Inputs		Outputs				
	Feedstock [a]	Electricity [b]	Water [a]	HEFA-SPK [a]	Diesel [a]	Naphtha [a]	Fuel Gas [a]
1 (reference)	500	988	222	227	40	135	26.4
2 (dynamic)	500	996	222	227	40	135	26.4
3 (stand-alone)	500	999	222	227	40	135	26.4
4 (natural gas)	500	33	254	227	40	135	36.2
5 (biomethane)	500	33	254	227	40	135	36.2
6 (naphtha)	500	32	279	227	40	88	13.7
7 (rapeseed)	500	1088	240	237	45	123	27.1
8 (palm)	500	884	203	249	104	38	25.1
9 (UCO)	500	1096	241	235	43	129	24.7
10 (diesel)	500	826	192	60	328	22	14.7

[a] $kt\,a^{-1}$; [b] $GWh\,a^{-1}$.

The reference scenario requires an input of 500 kt a^{-1} jatropha oil. Assuming a yield of established jatropha plantations of 0.5 to 2.2 t ha^{-1} a^{-1} [45–47], a cultivation area of between 0.227 and 0.833 million ha is necessary, which would constitute a substantial portion of the currently cultivated area of roughly 1 million ha. In order to produce the required 999 GWh a^{-1} of electricity in the stand-alone scenario using 100% wind and solar power (scenario 3), an appropriate combination would be a wind farm with 360 MW installed power and a solar power park with an area of 625 hectares. In comparison, this is 0.7% of the currently installed wind power in Germany and is substantially larger than the currently largest German solar park of 363 hectares.

In the scenarios with dynamic power procurement and with a stand-alone power production (scenarios 2 and 3), large storage capacities are required (775 t and 1585 t of hydrogen). The most suitable solution is a cavern storage facility in underground salt layers, which can be found in north and central Germany. Since the feedstock will be transported over high seas (except rapeseed and UCO), and with offshore wind farms being more productive, a coastal region close to a large international airport is most suitable for a PTG-HEFA plant of the described dimensions.

Due to its favorable composition, palm oil requires the least hydrogen when being processed into HEFA-SPK. From a purely technical point of view, palm oil would therefore be favorable when hydrogen is being produced by electrolysis. In the scenarios 4–6, where hydrogen production via steam reforming is analysed, the electricity demand is much lower than all other scenarios where the electricity demand derives almost entirely from the energy demand of the electrolysis. Especially in scenario 6, where internally produced naphtha and fuel gas are sufficient for the steam reforming of the required hydrogen, this is a preferable alternative to the reference scenario. Considering the requirements for hydrogen storage and electricity production in the stand-alone scenario, this is especially apparent.

3.2. Environmental Assessment

The GHG balances for the scenarios have been conducted according to the described methodology. Referring to Figure 3, the provision of jatropha-based HEFA-SPK can achieve a GHG reduction of 76% compared to the fossil reference defined in RED. Prerequisite is the exclusive use of electricity from wind and solar systems to cover the process energy demand. The provision of hydrogen via electrolysis is electricity-intensive and the use of grid electricity under the forecasted conditions leads to significantly higher emissions. In particular, the relatively high share of fossil fuels in the German electricity mix prevents higher GHG reductions. Other ways to reduce emissions from the supply of HEFA-SPK, in addition to the use of renewable electricity or a significantly higher share of the same in the electricity mix, is the use of in-process provided naphtha for hydrogen production via steam reforming. Reductions of more than 70% can be achieved here. On the other hand, the use of sustainably grown jatropha, which means jatropha grown mainly from extensive cultivation, can reduce overall GHG emissions.

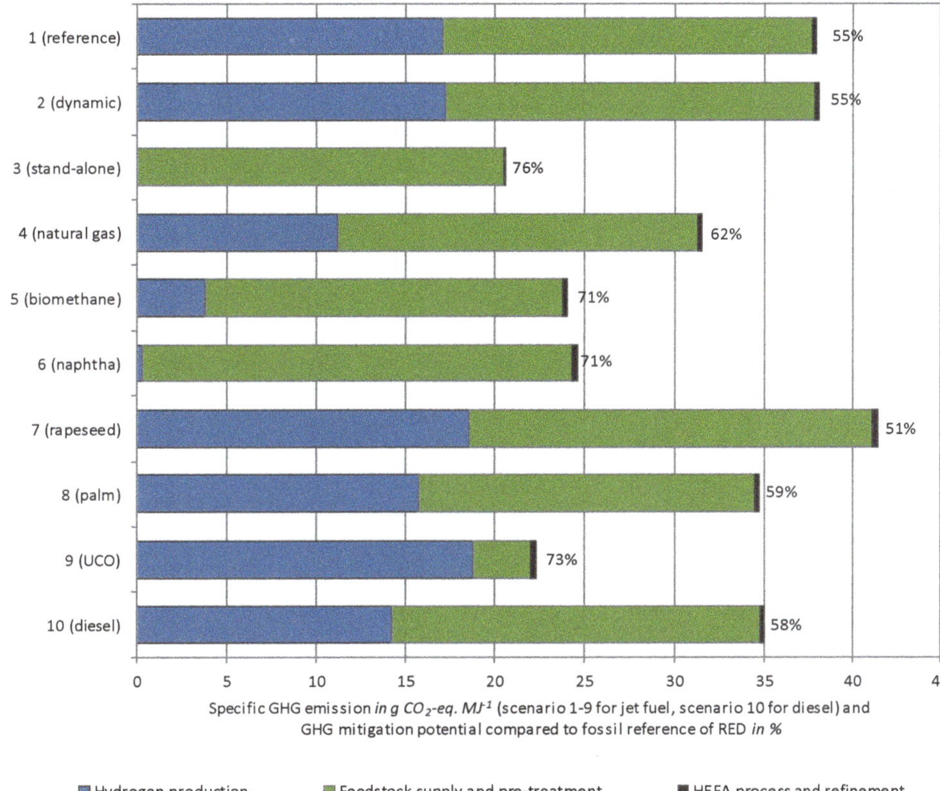

Figure 3. Specific GHG emission per scenario for HEFA-SPK and HEFA-diesel, and GHG mitigation potential compared to fossil reference of the European Renewable Energy Directive (RED).

3.3. Economic Assessment

Based on the results of the technical assessment, specific production costs were calculated. Referring to Figure 4, the bandwidth of total production costs is from 1295 to 1800 EUR t^{-1} for HEFA-SPK, which is well above the current market price of fossil jet fuel of around 425 EUR t^{-1}. Costs can be lowered to 1210 EUR t^{-1} for shifting the product values to diesel; this is due to the smaller production of short-chained, low-value naphtha and fuel gases.

When analysing the impact of the specific cost components to the total costs, feedstock costs dominate and the contribution from the PTG part is significant. Costs of capital and operation seem negligible in that context. The scenario of steam reforming with naphtha and fuel seems to be more favorable than PTG in terms of costs. This is due to the fact that only small amounts of revenue are lost but large investments in electrolysers and hydrogen storage and therefore also electricity costs are saved.

A more detailed discussion of the results, an exemplarily sensitivity analysis and comparisons with results of similar investigations for HEFA fuels are presented in [12].

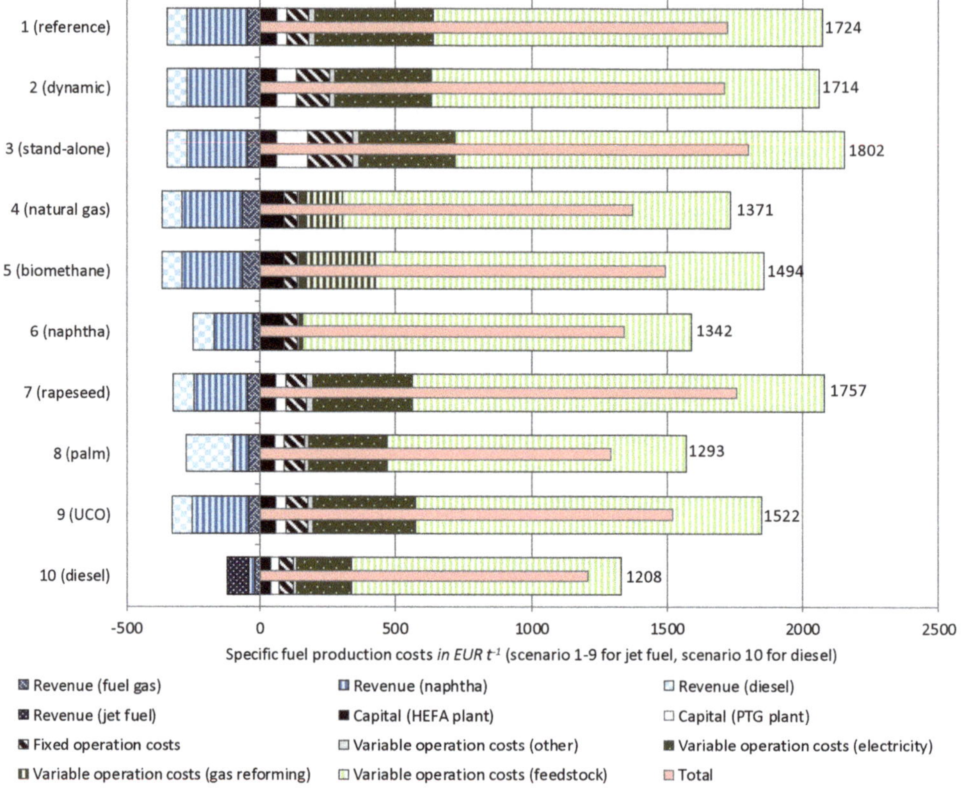

Figure 4. Specific fuel production costs per scenario.

3.4. Excursion on Favorable Regions

The results of the GHG and cost calculations are shown in Figure 5, with indication of the minimum GHG mitigation potential according to RED compared to the fossil reference (60% correspond to 33.8 g CO_2 eq MJ^{-1}) and the price for fossil jet fuel (about 425 EUR t^{-1}). The most favorable scenarios are the ones with the lowest production costs and at the same time lowest GHG emissions, i.e. with the lowest GHG mitigation costs. According to this, scenario 6 with internal use of naphtha for hydrogen supply might be the most favorable option, followed by scenario 10 if diesel is the main product and scenario 8 if palm oil is used as feedstock. Additionally, the exemplary cases for favorable regions do not show many benefits compared to the German scenarios. Out of them, the reference case in Sweden and the use of wind power for hydrogen production in Spain are the most favorable options.

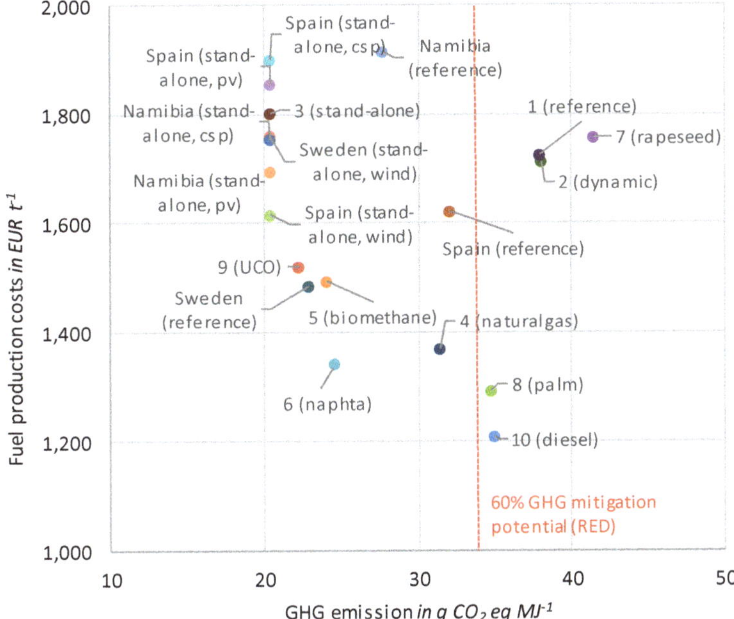

Figure 5. Comparison of production costs and GHG emissions for different scenarios and exemplary favorable regions.

4. Conclusions

Against the background of a growing aviation sector and the urgent demand on very large amounts of sustainable aviation fuels (SAF), an approach combining existing HEFA technology together with alternative hydrogen supply for processing realized via renewable PTG has been analysed in a comprehensive feasibility study. Ten different scenarios with regard to electricity supply for the PTG part, the hydrogen supply via electrolysis or steam reforming, different feedstocks for the HEFA part, as well as different product mixes were investigated from technical, economic, and environmental points of view.

The technical assessment shows the specific mass and energy balances for the scenario-specific designs of such PTG-HEFA hybrid refineries. They differ, for instance, with regard to hydrogen demand, process energy and related product streams. The least hydrogen is required when processing palm oil to HEFA-SPK (scenario 8) and from a technical point of view it is favorable to generate hydrogen by steam-reforming, as the electrolysis causes the predominant part of the electricity demand of the plant and, in the dynamic and stand-alone scenario (2 and 3), creates technical challenges for storing the hydrogen that can be avoided with constant hydrogen production.

The environmental assessment results allow one to conclude that for all scenarios at least 50% GHG mitigation can be achieved; but following the EU RED with 60% for new plants, just five of 10 concepts achieve that requirement. The use of as much renewable electricity as possible for the PTG part, integrated or other biogenic process energy, and the choice of feedstocks are the most relevant options for high GHG mitigation potentials. Accordingly, all stand-alone scenarios, steam reforming with naphtha from the HEFA plant, or steam reforming with biomethane and the usage of UCO as feedstock are environmentally more favorable.

The economic assessment reveals that the specific fuel production costs highly depend on feedstocks and PTG-related costs. Total production costs are several times higher compared to the current market price of jet fuel. Disregarding scenario 10 where diesel is the main product, the lowest

HEFA-SPK production costs are offered, in order, by the scenarios where palm oil is used as feedstock or hydrogen is produced by steam reforming of internal naphtha or natural gas.

Furthermore, the analysis done regarding favorable regions for the PTG part does not show significant benefits of one scenario over the others, as all have very low GHG emissions but are in the upper range of the price spectrum of all examined scenarios. Although producing slightly higher GHG emissions than the stand-alone options, placing the reference scenario in Sweden would result in the most favorable option as the production costs are the lowest when considering different regions.

Taking all investigated scenarios and assessments into account, the lowest GHG mitigation costs arise when internally produced naphtha is used for the hydrogen production by steam reforming (scenario 6). Focussing on diesel as the main product (10) causes just slightly higher GHG mitigation costs but does not offer the required GHG mitigation potential of 60% or more. The next best option in terms of mitigation costs, placing the reference scenario in Sweden, has, compared with the previous two options, the highest GHG mitigation potential at 73%.

The limited HEFA capacities can cover only a small part of the energy demand of the world aviation sector, and all scenarios show higher costs than current kerosene supply does. Although other technologies for renewable sustainable aviation fuels are under R&D&D and new aviation technologies might become relevant in the long-term, the same applies to aviation and all other transport sectors as well: total energy demand needs to be drastically reduced and SAF have to be applied in parallel.

Author Contributions: F.M.-L.: structured and drafted most parts of this article (writing—original draft preparation) based on the results of the above mentioned project which she were the project coordinator and did funding acquisition. K.O.: methodology (environmental assessment), resources, data curation and formal analysis (environmental assessment), writing—review and editing the paper. S.D.: methodology (technical assessment of HEFA part), resources, data curation and formal analysis (technical assessment: calculation of mass and energy balances and definition of infrastructural requirements), writing—review and editing of the paper. K.M.Z.: methodology (economic assessment), resources, data curation and formal analysis (economic assessment), writing, internal review of the paper. M.R.: resources, data curation and formal analysis of electricity supply scenarios in Germany and other countries (esp. demand, capacities, costs), internal review of the paper. W.W.: resources, data curation and formal analysis of technical issues and costs of the PTG plant, internal review of the paper.

Funding: This research was funded by the German Federal Ministry of Transport and Digital Infrastructure (BMVI) and was part of the collaborative project "Feasibility study on a PTG-HEFA hybrid refinery", investigated from 2015-2017 [3]. The APC was funded by BMVI as well.

Conflicts of Interest: The authors declare no conflict of interest.

Abbreviations

a	annum/year
aireg	Aviation Initiative for Renewable Energy in Germany e.V.
app.	approximately
ASTM	American Society for Testing and Materials
CAP	Climate action plan 2050 for Germany
csp	concentrated solar power
EU	European Union
GHG	greenhouse gas emissions
GWP	global warming potential
HEFA	hydrotreated esters and fatty acids
IATA	International Air Transport Association
kt	kilo tons (1000 t)
LCA	Life Cycle Assessment
PJ	Petajoule
PTG	power-to-gas (here: based on electrolysis to hydrogen)
PTL	power-to-liquid
pv	photovoltaics
RED	Renewable Energy Directive

SAF	sustainable aviation fuels
SPK	synthetic paraffinic kerosine
SynBioPTx	synergies (syn) of biomass-based (bio) and electricity-based (PTx) fuels and product processes
UCO	used cooking oil
VDI	Verein Deutscher Ingenieure (Association of German Engineers')

References

1. EASA. *European Aviation Environmental Report 2019*; EASA, EEA, Eurocontrol: Brussels, Belgium, 2019. [CrossRef]
2. Le Feuvre, P. Commentary: Are Aviation Biofuels Ready for Take Off? Available online: https://www.iea.org/newsroom/news/2019/march/are-aviation-biofuels-ready-for-take-off.html (accessed on 22 March 2019).
3. Dietrich, S.; Oehmichen, K.; Zech, K.; Müller-Langer, F.; Majer, S.; Kalcher, J.; Naumann, K.; Wirkner, R.; Pujan, R.; Braune, M.; et al. Studie im Rahmen der Mobilitäts-und Kraftstoffstrategie der Bundesregierung (MKS) im Auftrag für das Bundesministerium für Verkehr und digitale Infrastruktur (BMVI). In *Machbarkeitsanalyse für eine PTG-HEFA Hybridraffinerie in Deutschland*; Deutsches Biomasseforschungszentrum gemeinnützige GmbH (DBFZ): Leipzig, Germany, 2017.
4. Naumann, K.; Schröder, J.; Oehmichen, K.; Etzold, H.; Müller-Langer, F.; Remmele, E.; Thuneke, K.; Raksha, T.; Schmidt, P. *Monitoring Biokraftstoffsektor*, 4th ed.; Deutsches Biomasseforschungszentrum Gemeinnützige GmbH (DBFZ): Leipzig, Germany, 2019; ISBN 978-3-946629-36-8.
5. Müller-Langer, F.; Dietrich, R.U.; Krol, R.V.D.; Arnold, K.; Harnisch, F.; Erneuerbare Kraftstoffe für Mobilität und Industrie. FVEE Themen 2016—Netze und Speicher für die Energiewende—Erneuerbare Kraftstoffe. 2016. Available online: http://www.fvee.de/fileadmin/publikationen/Themenhefte/th2016/th2016_07_05.pdf (accessed on 13 September 2019).
6. Müller-Langer, F.; Etzold, H.; Naumann, K. BTx and PTx as competitors or companions: A systemic assessment. In Proceedings of the 8th ETIP Stakeholder Plenary Meeting, Brussels, Belgium, 11–12 April 2018. Available online: http://www.etipbioenergy.eu/images/SPM8_Presentations/ETIP_Mueller-Langer_2018-04_new.pdf (accessed on 13 September 2019).
7. Schmidt, P.; Weindorf, W.; Roth, A.; Batteiger, V.; Riegel, F. *Power-to-Liquids—Potentials and Perspectives for the Future Supply of Renewable Aviation Fuel*; German Environment Agency: Dessau-Roßlau, Germany, 2016. Available online: https://www.umweltbundesamt.de/sites/default/files/medien/377/publikationen/161005_uba_hintergrund_ptl_barrierrefrei.pdf (accessed on 13 September 2019).
8. Schmidt, P.; Weindorf, W.; Zittel, W. Renewables in Transport 2050—Empowering a Sustainable Mobility Future with Zero Emission Fuels from Renewable Electricity—Europe and Germany. In *Report 1086–2016*; Forschungsvereinigung Verbrennungskraftmaschinen e.V.: Frankfurt am Main, Germany, 2015. Available online: http://www.lbst.de/news/2016_docs/FVV_H1086_Renewables-in-Transport-2050-Kraftstoffstudie_II.pdf (accessed on 13 September 2019).
9. Schmied, M.; Wüthrich, P.; Zah, R.; Althaus, H.J.; Friedl, C. Postfossile Energieversorgungsoptionen für einen treibhausgasneutralen Verkehr im Jahr 2050: Eine verkehrsträgerübergreifende Bewertung. In *TEXTE 30/2015—Umweltforschungsplan des Bundesministeriums für Umwelt, Naturschutz, Bau und Reaktorsicherheit*; German Environment Agency: Dessau-Roßlau, Germany, 2015. Available online: https://www.umweltbundesamt.de/sites/default/files/medien/378/publikationen/texte_30_2015_postfossile_energieversorgungsoptionen.pdf (accessed on 13 September 2019).
10. Albrecht, F.G.; Dietrich, R.-U. Alternative fuels from Biomass and Power (PBtL) A case study on process options, technical potentials, fuel costs and ecological performance. In Proceedings of the European Biomass Conference, Stockholm, Sweden, 12–15 July 2017. Available online: https://elib.dlr.de/115071/ (accessed on 13 September 2019).
11. Müller-Langer, F.; Vogel, A.; Brauner, S. *Renwew—Renewable fuels for advanced powertrains—Deliverable 5.3.8. Overall Costs*; Institute for Energy and Environment: Leipzig, Germany, 2008. Available online: http://www.renew-fuel.com/download.php?dl=del_sp5_wp3_5-3-8_08-02-27_iee-draft.pdf&kat=18 (accessed on 13 September 2019).
12. Zech, K.M.; Dietrich, S.; Reichmuth, M.; Weindorf, W.; Müller-Langer, F. Techno-economic assessment of a renewable bio-jet-fuel production using power-to-gas. *Appl. Energy* **2018**, *231*, 997–1006. [CrossRef]

13. Jauslin Stebler, AG. Erdgas-Röhrenspeicher Urdorf. Available online: https://www.jauslinstebler.ch/VGA/VEM/projekte/erdgas-roehrenspeicher-urdorf.html (accessed on 23 July 2018).
14. Doradei, S.; Crotogino, F.; Acht, A.; Horvarth, P.-L. Speicherung von Wasserstoff in Salzkavernen. In *Integration von Wind-Wasserstoff-Systemen in das Energiesystem*; Nationales Innovationsprogramm Wasserstoff-und Brennstoffzellentechnologie (NIP): Berlin, Germany, 2013.
15. Gröngröft, A.; Meisel, K.; Hauschild, S.; Grasemann, E.; Peetz, D.; Mayer, K.; Teil, I.I. Wissenschaftliche Untersuchung von Wegen der Biokerosinproduktion aus verschiedenen Biomassetypen. In *Abschlussbericht zu dem Vorhaben Projekt BurnFAIR*; Zschocke, A., Ed.; Deutsche Lufthansa: Frankfurt am Main, Germany, 2014.
16. Liu, Y.; Sotelo-Boyás, R.; Murata, K.; Minowa, T.; Sakanishi, K. Hydrotreatment of vegetable oils to produce bio-hydrogenated diesel and liquefied petroleum gas fuel over catalysts containing sulfided Ni–Mo and solid acids. *Energy Fuels* **2011**, *25*, 4675–4685. [CrossRef]
17. Dubois, V.; Breton, S.; Linder, M.; Fanni, J.; Parmentier, M. Fatty acid profiles of 80 vegetable oils with regard to their nutritional potential. *Eur. J. Lipid Sci. Technol.* **2007**, *109*, 710–732. [CrossRef]
18. Abidin, S.Z.; Patel, D.; Saha, B. Quantitative analysis of fatty acids composition in the used cooking oil (UCO) by gas chromatography–mass spectrometry (GC-MS). *Can. J. Chem. Eng.* **2013**, *91*, 1896–1903. [CrossRef]
19. Majer, S.; Hauschild, S.; Müller-Langer, F. Kurzstudie im Auftrag des Verbandes der Deutschen Biokraftstoffindustrie e.V., der Union zur Förderung von Öl-und Proteinpflanzen e.V. und des OVID Verband der ölsaatenverarbeitenden Industrie in Deutschland e.V. In *Energie-und Treibhausgasbilanz von HVO-Kraftstoff. Eine vergleichende Analyse*; Deutsches Biomasseforschungszentrum gemeinnützige GmbH (DBFZ): Leipzig, Germany, 2014.
20. Endisch, M.; Balfanz, U.; Olschar, M.; Kuchling, T. Vegetable Oil Hydrotreating for Production of High Quality Diesel Components. In Proceedings of the DGMK Future Feedstocks for Fuels and Chemicals Conference, Berlin, Germany, 29 September–1 October 2008.
21. Nikander, S. Greenhouse Gas and Energy Intensity of Product Chain: Case Transport Bio-fuel. Master's Thesis, Helsinki University of Technology, Helsinki, Finland, 9 May 2008.
22. Smejkal, Q.; Smejkalová, L.; Kubička, D. Thermodynamic balance in reaction system of total vegetable oil hydrogenation. *Chem. Eng. J.* **2008**. [CrossRef]
23. Jęczmionek, Ł.; Porzycka-Semczuk, K. Hydrodeoxygenation, decarboxylation and decarbonylation reactions while co-processing vegetable oils over NiMo hydrotreatment catalyst. Part II. Thermal effects—Experimental results. *Fuel* **2014**, *128*, 296–301. [CrossRef]
24. Hiller, H.; Reimert, R.; Stönner, H.-M. Gas Production. 1. Introduction. In *Ullmann's Encyclopedia of Industrial Chemistry*; Wiley-VCH: Weinheim, Germany, 2014. [CrossRef]
25. Kinder, J.D.; Rahmes, T. Evaluation of Bio-Derived Synthetic Paraffinic Kerosene (Bio-SPK). 2009. Available online: http://www.safug.org/assets/docs/biofuel-testing-summary.pdf (accessed on 15 August 2019).
26. Myllyoja, J.; Aalto, P.; Savolainen, P.; Purola, V.-M.; Alopaeus, V.; Grönqvist, J. Process for the Manufacture of Diesel Range Hydrocarbons. U.S. Patent US8022258 B2, 20 September 2011.
27. German Institute for Standardisation. *Environmental Management—Life Cycle Assessment—Principles and Framework*; DIN ISO 14040: Geneva, Switzerland, 2006.
28. German Institute for Standardisation. *Environmental Management—Life Cycle Assessment—Requirements and Guidelines*; DIN ISO 14044: Geneva, Switzerland, 2006.
29. European Commission. Directive 2009/28/EC of the European Parliament and of the Council of 23 April 2009 on the promotion of the use of energy from renewable sources and amending and subsequently repealing Directives 2001/77/EC and 2003/30/EC. *Off. J. Eur. Union* **2009**. Available online: http://www.nezeh.eu/assets/media/fckuploads/file/Legislation/RED_23April2009.pdf (accessed on 27 September 2019)
30. Ecoinvent Center. *Ecoinvent Version 3 Life Cycle Inventory Database*; Swiss Center for Life Cycle Inventories: St. Gallen, Switzerland, 2016; database.
31. IINAS GmbH. GEMIS—Global Emissions Model for integrated Systems V4.94. Version: International Institute for Sustainability Analysis and Strategy GmbH. Database. Available online: http://www.gemis.de (accessed on 26 September 2019).
32. Solomon, S.; Quin, D.; Manning, M.; Chen, Z.; Marquis, M.; Averyt, K.B. *Climate Change 2007: The Physical Science Basis. Contribution of Working Group I to the Fourth Assessment Report of the Intergovernmental Panel on Climate Change*; Cambridge University Press, IPCC: Cambridge, UK; New York, NY, USA, 2007.

33. The Association of German Engineers (VDI). *Guideline 6025—Economy Calculation Systems for Capital Goods and Plants*; Beuth-Verlag GmbH: Berlin, Germany, 2012.

34. National Renewable Energy Laboratories (NREL). Task 1: Cost estimates of small modular systems. In *Equipment Design and Cost Estimation for Small Modular Biomass Systems, Synthesis Gas Cleanup, and Oxygen Separation Equipment*; NREL/SR-510-39943; National Renewable Energy Laboratories (NREL): Golden, CO, USA, 2006.

35. Davis, R.; Kinchin, C.; Markham, J.; Tan, E.; Laurens, L.; Sexton, D.; Knorr, D.; Schoen, P.; Lukas, J. *Process Design and Economics for the Conversion of Algal Biomass to Biofuels: Algal Biomass Fractionation to Lipid and Carbohydrate-Derived Fuel Products*; NREL/TP-5100-62368; National Renewable Energy Laboratories (NREL): Golden, CO, USA, 2014. Available online: https://www.nrel.gov/docs/fy14osti/62368.pdf (accessed on 3 September 2019).

36. Eurostat. Harmonized index for consumer prices; industrial goods; Table: [prc_hicp_aind]; 2016. Available online: http://appsso.eurostat.ec.europa.eu/nui/show.do?dataset=prc_hicp_aind&lang=de (accessed on 25 September 2019).

37. Eurostat. Euro/Ecu exchange rates; Table: [ert_bil_eur_a]; 2018. Available online: http://appsso.eurostat.ec.europa.eu/nui/show.do?dataset=ert_bil_eur_a&lang=en (accessed on 25 September 2019).

38. Deutsches Zentrum für Luft-und Raumfahrt (DLR); Ludwig-Bölkow-Systemtechnik (LBST); Fraunhofer ISE; KBB Underground Technologies. Studie über die Planung einer Demonstrationsanlage zur Wasserstoff-Kraftstoffgewinnung durch Elektrolyse mit Zwischenspeicherung in Salzkavernen unter Druck. Stuttgart, Germany, 2015. Available online: http://www.lbst.de/ressources/docs2015/BMBF_0325501_PlanDelyKaD-Studie.pdf (accessed on 23 July 2018).

39. Nationales Innovationsprogramm Wasserstoff-und Brennstoffzellentechnologie (NIP). Integration von Wind-Wasserstoff-Systemen in das Energiesystem, Berlin, 2013. Available online: https://www.now-gmbh.de/content/1-aktuelles/1-presse/20140402-abschlussbericht-zur-integration-von-wind-wasserstoff-systemen-in-das-energiesystem-ist-veroeffentlicht/abschlussbericht_integration_von_wind-wasserstoff-systemen_in_das_energiesystem.pdf (accessed on 18 July 2018).

40. Braune, M.; Grasemann, E.; Gröngröft, A.; Klemm, M.; Oehmichen, K.; Zech, K. *Die Biokraftstoffproduktion in Deutschland—Stand der Technik und Optimierungsansätze*, 1st ed.; DBFZ report no. 22; Deutsches Biomasseforschungszentrum gemeinnützige GmbH (DBFZ): Leipzig, Germany, 2016; ISBN 978-3-9817707-8-0.

41. Turton, R.; Shaeiwitz, J.A.; Bhattacharyya, D.; Whiting, W.B. *Analysis, Synthesis and Design of Chemical Processes*, 5th ed.; Pearson Education; Prentice Hall International Series in the Physical and Chemical Engineering Sciences: Boston, MA, USA, 2018.

42. Aviation Initiative for Renewable Energy in Germany e. V. (aireg). Database on Jatropha. 2015; unpublished.

43. Adler, P.; Billig, E.; Brosowski, A.; Daniel-Gromke, J.; Falke, I.; Fischer, E.; Grope, J.; Holzhammer, U.; Postel, J.; Schnutenhaus, J.; et al. (Eds.) *Leitfaden Biogasaufbereitung und-einspeisung*; Fachagentur Nachwachsende Rohstoffe e.V.: Gülzow, Germany, 2014; ISBN 3-00-018346-9.

44. Oil Price Information Service (OPIS). *Europe Jet, Diesel and Gasoil Report*; Oil Price Information Service (OPIS): Rockville, MD, USA, 2016.

45. Kumar, S.; Singh, J.; Nanoti, S.M.; Garg, M.O. A comprehensive life cycle assessment (LCA) of Jatropha biodiesel production in India. *Bioresour. Technol.* **2012**, *110*, 723–729. [CrossRef] [PubMed]

46. Eshton, B.; Katima, J.H.Y.; Kituyi, E. Greenhouse gas emissions and energy balances of jatropha biodiesel as an alternative fuel in Tanzania. *Biomass Bioenergy.* **2013**, *58*, 95–103. [CrossRef]

47. Dehue, B.; Hettinga, W. GHG Performance Jatropha Biodiesel. Commissioned by D1 oils, Ref no. PBIONL073010. Ecofys Reference 2. 2008. Available online: https://pdfs.semanticscholar.org/2f0b/c772fedd5e7988593084bd4113c1a5c78472.pdf (accessed on 26 September 2019).

 *applied sciences*

Review

What is still Limiting the Deployment of Cellulosic Ethanol? Analysis of the Current Status of the Sector

Monica Padella *, Adrian O'Connell and Matteo Prussi

European Commission, Joint Research Centre, Directorate C-Energy, Transport and Climate, Energy Efficiency and Renewables Unit C.2-Via E. Fermi 2749, 21027 Ispra, Italy; adrian.oconnell@ec.europa.eu (A.O.); matteo.prussi@ec.europa.eu (M.P.)
* Correspondence: monica.padella@ec.europa.eu

Received: 16 September 2019; Accepted: 15 October 2019; Published: 24 October 2019

Abstract: Ethanol production from cellulosic material is considered one of the most promising options for future biofuel production contributing to both the energy diversification and decarbonization of the transport sector, especially where electricity is not a viable option (e.g., aviation). Compared to conventional (or first generation) ethanol production from food and feed crops (mainly sugar and starch based crops), cellulosic (or second generation) ethanol provides better performance in terms of greenhouse gas (GHG) emissions savings and low risk of direct and indirect land-use change. However, despite the policy support (in terms of targets) and significant R&D funding in the last decade (both in EU and outside the EU), cellulosic ethanol production appears to be still limited. The paper provides a comprehensive overview of the status of cellulosic ethanol production in EU and outside EU, reviewing available literature and highlighting technical and non-technical barriers that still limit its production at commercial scale. The review shows that the cellulosic ethanol sector appears to be still stagnating, characterized by technical difficulties as well as high production costs. Competitiveness issues, against standard starch based ethanol, are evident considering many commercial scale cellulosic ethanol plants appear to be currently in idle or on-hold states.

Keywords: cellulosic ethanol; GHG savings; R&D funding

1. Introduction

Ethanol production from cellulosic material such as agricultural residues (e.g., wheat straw and corn stover) and energy crops (e.g., switchgrass and miscanthus) is considered a highly promising option for future ethanol production, helping the energy diversification and decarbonization of the transport sector. Compared to conventional (or first generation) ethanol production from food and feed crops (mainly sugar and starch based crops), cellulosic (or second generation, 2G) ethanol is considered to provide better performance in terms of greenhouse gas (GHG) reduction and low risk of direct and indirect land-use change impacts. Those advantages have led to the promotion of cellulosic ethanol in the legislation around the globe.

In the EU, the current 2009 Renewable Energy Directive (RED) [1] and Fuel Quality Directive (FQD) [2] (in force until 2020), contain a 10% renewable energy target for the transport sector by 2020 and a 6% GHG reduction target for transport (set in FQD) that are both expected to be met largely with food-based biofuels. In response to the controversial issue of indirect land use change (ILUC), related to the global market-mediated agricultural area expansion, the 2015 ILUC Directive [3], amending RED and FQD, introduced a cap on crop-based biofuels, which can contribute up to 7% of the final consumption of transport energy in 2020. The directive [3] also introduced requirements for reporting ILUC emissions and measures to promote advanced biofuels. It [3] requires Member States to promote the use biofuels produced from feedstocks listed in Part A of Annex IX (which includes waste, residue, non-food cellulosic, and ligno-cellulosic feedstocks) by setting a non-legally binding target of 0.5% in energy content for their use in transport fuel. RED [1] also has a 'double-counting' mechanism, according to which the energy content of those biofuels is double counted towards the overall renewable energy in transport target. The recast of RED, Directive EU 2018/2001 [4], with effect from 1 July 2021, makes advanced biofuel mandate compulsory containing a sub-target of 3.5% for advanced biofuels for 2030 within the 14% target for renewable energy in transport. Advanced biofuels are defined by the directive as 'biofuels that are produced from the feedstock listed in Part A of Annex IX' that includes among others, agriculture and forestry residues as well as energy crops. Those biofuels will continue to count double towards the targets. In addition, biofuels must meet a 65% greenhouse gas reduction threshold (in installation starting operation from 1 January 2021), to try to ensure substantial GHG savings compared to fossil-based fuels.

Focusing on the US, a fuel is classified as advanced on the basis of its lifecycle GHG savings (which need to be at least 50% when compared to fossil fuels). Cellulosic biofuel is an advanced biofuel subcategory: it is required to have a greater than or equal to 60% GHG saving compared to a 2005 petroleum baseline [5]. The Renewable Fuel Standard (RFS) program required 36 billion gallons of biofuels to be used by 2022, of which at least 16 billion gallons were supposed to come from cellulosic biofuels [6]. However, every year, the Environmental Protection Agency (EPA) has exercised its 'cellulosic waiver authority' to reduce the cellulosic biofuel target. In 2014, EPA also redefined the term 'cellulosic biofuels' expanding the definition to include certain types of biogas and ethanol from corn kernel fiber [7]. Since then, most of the increase in cellulosic biofuel consumption is from biogas, in either compressed or liquefied form. The current cellulosic biofuel mandate (2019) has been set by EPA at 418 million gallons, corresponding to about 5% of the target envisioned by the Energy Independence and Security Act of 2007 [8]. Moving to Brazil, the recent RenovaBio program aims to decrease transport emissions by 10% in the next 10 years favoring fuels with lower carbon intensities [9]. Like the US Renewable Fuel Standard, it favors fuels with lower carbon intensities and biofuel producers will receive credits based on the lifecycle emission savings of their fuel compared to fossil fuel. Other programs for the promotion of second generation ethanol in Brazil include the Joint Support Plan for Industrial Technological Innovation in Sugarcane-based Ethanol and Chemistry Sectors (PAISS Program) that provides funding aiming to boost Brazil's presence in more advanced technologies [10].

In China, the "Implementation Plan for the Expansion of Ethanol Production and Promotion for Transportation Fuel", jointly announced by the National Development and Reform Commission (NDRC) and other ministries in September 2017, targets a national shift in production to cellulosic technology by 2025. Cellulosic ethanol production is supported via a subsidy of 600 RMB (Renminbi) per ton (about 78 €/t) since 2014. However, prospects for 2018 are uncertain since updates on the subsidy program have not been announced at this time [11].

In addition to the aforementioned policy measures, the evolution of the cellulosic ethanol sector is influenced by various aspects, such as past and current R&D investments, production costs as well as technical and environmental issues that must all be taken into account in order to provide a comprehensive picture of the sector and its future challenges.

Already in 2002, Badger [12] provided a general review on ethanol from cellulose in US concluding that "although several ethanol-from-cellulose processes are technically feasible, cost-effective processes have been difficult to achieve". Kumar et al. in 2016 [13] and more recently Liu et al. in 2019 [14] reached similar conclusions. They both reported that cellulosic ethanol is still not competitive compared to conventional ethanol and efforts are still needed to reduce costs.

The main purpose of this review is to provide a comprehensive overview of the status and different aspects of cellulosic ethanol production both inside and outside the EU, highlighting technical and non-technical barriers that still limit its production at commercial scale. A literature review of the status of cellulosic ethanol plants in the EU and worldwide is carried out in Section 2. Section 3 reports GHG emission savings and production costs associated to cellulosic ethanol production.

An analysis of R&D funding and programs mainly in EU is provided in Section 4, while barriers to large scale deployment of cellulosic ethanol are discussed in Section 5. Section 6 draws conclusions.

2. Current Cellulosic Ethanol Production Process in Commercial Plants

Cellulosic ethanol production can be summarized in four main steps: pretreatment, hydrolysis, fermentation, and product recovery (Figure 1).

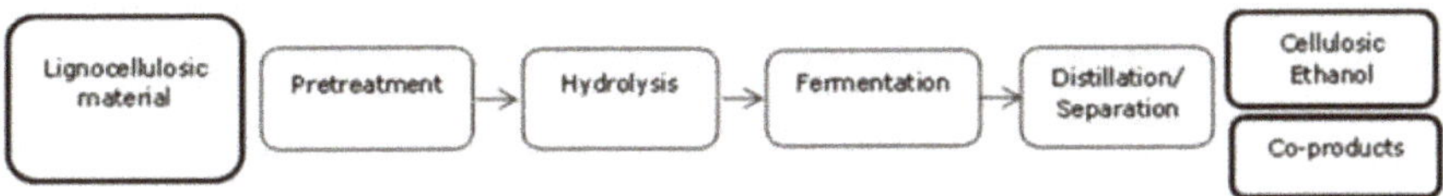

Figure 1. Cellulosic ethanol process.

Pretreatment makes the lignocellulosic biomass more amenable to biological conversion ensuring complete substrate utilization. Several pretreatment technologies have been developed and tested at various scales [13,14], and steam explosion has been recognized as the one most widely used by industrial companies [15]. In addition, hydrolysis (of the cellulosic and hemi-cellulosic structures) can also be enhanced by means of enzymes or dilute acid pretreatments. Following pretreatment, enzymatic hydrolysis is the most commonly applied hydrolysis method, although the cost of enzymes currently takes a major proportion of the production costs [15] (see Section 3 for more details on production costs). Bacteria or yeasts are then used to ferment sugars into ethanol.

In terms of production, around 109 billion liters (or 86 million tons) of ethanol were globally produced in 2018, most of which was made by US (56%), followed by Brazil (28%), EU (5%), and China (4%) [16]. For cellulosic ethanol, commercial size plants have been constructed in Europe, US, Brazil, and China, but regular and reliable production is yet to be proven. The actual cellulosic ethanol production to date has been markedly below the installed capacity.

For the EU, ePURE reports that almost 200 kt of European ethanol has been produced from ligno-cellulosic, other RED Annex IX-A feedstocks, or other feedstocks, representing 4% of the total ethanol production in 2017 [17]. However, according to the US Department of Agriculture (USDA) Global Agricultural Information Network (GAIN) report, EU28 annual production of cellulosic ethanol was estimated to be around 40 kt in 2017 down to 10 kt in 2018 [18]. In the US, the EPA's 2018 RFS data reports US cellulosic biofuel production levels, from which it's possible to estimate a total of about 30 kt of cellulosic ethanol produced domestically in 2017 [8]. In Brazil, total cellulosic ethanol production is estimated to be 25 million liters (or 20 kt) for 2018 representing an insignificant share of total ethanol production in Brazil [9]; while, for China, 2018 cellulosic ethanol production is forecast to stop at 20 million liters (or 16 kt) as its major cellulosic project appears idle [11].

There are several first-of-a-kind commercial scale cellulosic ethanol plants at a global level, according to the International Energy Agency (IEA) database [19] and other sources [9,11,18,20–24], although operations of some of the plants are currently idle or on-hold (Table 1).

Only a few commercial scale plants are reported as operational in Norway, US, and Brazil.

In addition to the plants listed in Table 1, Quad County Corn Processors and six other plants (using the Edeniq technology) adapted their conventional corn ethanol refineries to produce ethanol from corn kernel fiber, known as 1.5 generation technology. As reported in Section 1, it will qualify as cellulosic biofuel in US following the EPA definition. However, corn kernels consist of a significant content of fibers (10–12%) that impact on the conversion of fermentable sugar, but they are mainly comprised of starch [7]; therefore, they should not be strictly considered as second generation ethanol production plants. Other plants are also planning to produce 1.5 generation ethanol: D3Max and ICM are developing their own technologies [7].

Table 1. Overview on a global commercial scale cellulosic ethanol plants.

Company	Project	Country	Output Capacity (ktons)	Status	Start-Up Year
Abengoa Bioenergy Biomass of Kansas, LLC	Commercial (acquired by Synata Bio Inc. [21])	US	75	idle	2014
Aemetis	Aemetis Commercial	US	35	planned	2019
Beta Renewables (acquired by Versalis [22])	Alpha	US	60	on hold	2018
Beta Renewables (acquired by Versalis)	Energochemica	EU (Slovakia)	55	on hold	2017
Beta Renewables (acquired by Versalis)	Fujiang Bioproject	China	90	on hold	2018
Beta Renewables [1] (acquired by Versalis)	IBP-Italian Bio Fuel	EU (Italy)	40	idle	2013
Borregaard Industries AS	ChemCell Ethanol	Norway	16	operational	1938
Clariant	Clariant Romania	EU (Romania)	50	under construction	2020
COFCO Zhaodong Co.	COFCO Commercial	China	50	planned	2018
DuPont	Commercial facility Iowa (acquired by VERBIO [23])	US	83	idle	2016
Enviral	Clariant Slovakia	EU (Slovakia)	50	planned	2021
Fiberight LLC	Commercial Plant	US	18	under construction	2019
GranBio	Bioflex 1	Brazil	65	operational	2014
Henan Tianguan Group	Henan 2	China	30	Idle	2011
Ineos Bio	Indian River County Facility (acquired by Alliance Bio-Products in 2016 [24])	US	24	idle	NA
Longlive Bio-technology Co. Ltd.	Longlive	China	60	Idle	2012
Maabjerg Energy Concept Consortium	Flagship integrated biorefinery	EU (Denmark)	50	on hold	2018
POET-DSM Advanced Biofuels	Project Liberty	US	75	operational	2014
Raízen Energia	Brazil	Brazil	36	operational	2015
St1 Biofuels Oy in cooperation with North European Bio Tech Oy	Cellunolix®	EU (Finland)	40	planned	2020

[1] Joint venture of Mossi & Ghisolfi Chemtex division with TPG.

The POET-DSM Advanced Biofuels LLC is reported as operational, but it is not clear how much ethanol it is currently producing. The plant has been inaugurated in 2014 in Iowa with a production capacity of 75 kt per year of cellulosic ethanol from corn stover and cob. In summer 2017, the company installed a new pretreatment technology and announced the construction of an on-site enzyme manufacturing facility that will cut costs associated with the process [25]. According to Schill [20], the plant has reached the targeted 70 gallons per ton of biomass and further optimization is in progress.

One of the world's largest cellulosic ethanol production facilities, the Beta Renewables plant officially opened at Crescentino (Italy) in 2013 but it has been shut down since 2017 due to a restructuring effort of the parent chemical company Mossi & Ghisolfi [26]. The plant had an annual capacity of 40 kt of ethanol produced from wheat straw, rice straw, and giant reed (Arundo donax). On November 1, 2018, Eni's chemical subsidiary Versalis acquired the Mossi Ghisolfi Group's green portfolio and it is currently in the process of defining an action plan to restart the activities of the Crescentino plant [18].

Two commercial scale plants are in operation in Brazil. One is the Bioflex 1 plant from Granbio that began production in September 2014, with a current production capacity of about 65 kt per year; the other one is Raízen's Costa Pinto Unit with an annual production capacity of 36 kt of ethanol from bagasse using a technology developed by Iogen Energy, a joint venture of Raízen and Iogen Corp. According to USDA GAIN Brazil report [9], this is the only one producing at relatively large scale.

Other projects have been announced for the production of cellulosic ethanol, mainly in the EU. The construction of a new full scale commercial cellulosic ethanol plant has been announced in 2017 by Enviral, the largest producer of bioethanol in Slovakia, that recently signed a license agreement to use Clariant's Sunliquid®technology [27]. The plant is planned to be integrated into the existing facilities at the Enviral's Leopoldov site (in Slovakia) producing 50 kt per year of ethanol from agricultural residues. Clariant has also announced on-going construction of the first large-scale commercial Sunliquid®plant for the production of cellulosic ethanol in Romania [28].

The Cellunolix®project (managed by St1 Biofuels Oy in cooperation with North European Bio Tech Oy) is planned to be operational in 2020 in Finland with an annual capacity of 40 kt. The plant will use saw dust and recycled wood as feedstock and will be located at UPM's Alholma industrial area [18].

In China, COFCO announced plans to build several 63 million liters (or 50 kt) capacity cellulosic fuel ethanol plants in future [11].

3. GHG Emissions and Production Costs of Cellulosic Ethanol

Cellulosic ethanol from feedstocks such as agricultural residues and energy crops is generally considered to be environmentally sustainable providing higher reduction of GHG emissions and zero or low indirect emissions from land use change compared to conventional (or first generation) ethanol production from food and feed crops.

Focusing on the framework after 2020, the RED recast [4] specifies that biofuels must meet a 65% greenhouse gas reduction threshold, compared to fossil fuels, in installations starting operation from 1 January 2021.

The methodology to calculate 'life-cycle' GHG emissions is set by the Directive in Annex V part C [4]. GHG emissions should be calculated by taking into account all emissions associated to all steps of the production and use of biofuels, from cultivation of raw materials, to processing into biofuels, and transport and distribution of all products. Emissions from carbon stock changes caused by direct land-use change, if they occurred, should also be taken into account. The total GHG emissions are referred to as 'direct emissions'. The Directive includes a list of default GHG emission values for the main biofuels that economic operators can use only if biofuels are produced without direct land use change.

The default value associated to cellulosic ethanol from wheat straw is 16 gCO_2eq/MJ [4] that, compared to the fossil fuel comparator, results in 83% GHG emission reduction. A similar value can be assumed for perennial grass (such as miscanthus and switchgrass) ethanol [29].

However, those values do not include accounting of GHG emissions associated with changes in the carbon stock of land resulting from indirect land use change (ILUC) and other indirect effects.

The ILUC issue refers to global market-mediated agricultural area expansion in response to increased biofuel demand. If crops grown on existing arable land are used to make biofuels and are diverted from food and feed production, then the gap in the food supply will be partly filled by the expansion of cropland, because of the necessity to replace the food production. This is referred to as indirect land use change [30]. Economic models are typically used to estimate global land use change consequences due to an increased biofuel demand (more details on how the economic models work and on alternative approaches to estimate ILUC can be found in [30]). ILUC has been estimated by numerous studies in the literature and in regulatory analyses; results show that ILUC emissions can be significant for food-based biofuels [3,30–34]. Nevertheless, non-food feedstocks may also have a land use change impact as well as other indirect emissions as a result of the displacement with existing uses [29,31].

The emissions from land use change calculated by the GLOBIOM (Global Biosphere Management Model) [31] associated to the increase in the demand of cereal straw based ethanol in the EU corresponds to 16 g CO_2eq/MJ of ethanol (assuming 20 years of amortization) in the case of unsustainable straw removal and 0 g CO_2eq/MJ when straw removal is no more than 33–50% of the total straw biomass (considered as the sustainable straw removal in the study [31]). These emissions result from the conversion of land for new cropland for extra cereals production and from the soil organic carbon impact of removing crop residues in unsustainable management [31].

For energy crops/perennial grasses, land use change emissions depends on which type of land is used for their production (existing agricultural land, unused land such as abandoned agricultural land, or high-carbon stock land such as forests) which will in turn be determined by what is more profitable for farmers to do compared to growing other crops or leaving the land uncultivated [35]. Emissions from land-use change estimated by the GLOBIOM model [31] for miscanthus and switchgrass Fisher-Tropsch biodiesel (a value in the same order of magnitude can be assumed for ethanol) produced in the EU are negative (-12 gCO$_2$eq/MJ), indicating that land use change from these crops actually reduces GHG emissions compared to a baseline case without them, mainly due to the increase in soil carbon stock where they are grown [29,31]. Most other modeling studies, reviewed in [36], found that energy crops are not likely to displace food and fiber crops on agricultural land, but would mostly be grown on abandoned agricultural land, cropland-pasture, and other unused land with low-carbon stocks, resulting in low or negative ILUC emissions. However, there is some uncertainty in the magnitude of energy crop ILUC emissions and additional considerations on the environmental risks of energy cropping as well as options for risk mitigation can be found in [35].

A lifecycle analysis looking at indirect effects, i.e., including emissions which would arise when replacing a material that has been diverted from its original use into fuel production was performed by [29] for a number of pathways, and included wheat straw to ethanol. For this pathway, the current usage of straw in heat and power generation and livestock bedding and feed, mushroom cultivation, and horticulture was considered (more detail of the analysis can be found in [29]). The authors estimated a value of 8 g CO$_2$eq/MJ (central estimate) of indirect emission for wheat straw ethanol.

Direct and indirect emissions for the two considered pathways are summarized in the following table (Table 2).

Table 2. Estimated greenhouse gas (GHG) emissions (direct and indirect) associated with cellulosic ethanol production from cereal straw and perennial crops.

	Direct Emissions from RED Recast, Annex V—Default Values [4] (g CO$_2$eq/MJ)	Indirect Emissions (ILUC Emissions from Valin et al., 2015 [31]) (g CO$_2$eq/MJ)	Indirect Emissions (Displacement Emissions from ICCT, 2017 [29]) (g CO$_2$eq/MJ)
Ethanol from wheat straw	16	16 if unsustainable straw removal is assumed (more than 33–50%) 0 if sustainable straw removal is assumed (less than 33–50%)	8 displaced uses: Livestock bedding and feed; mushroom cultivation; horticulture; heat and power
Ethanol from perennial grass (miscanthus and swithgrass)	16	-12	0 no existing uses

Several sources mention costs as the main issue for the development of the cellulosic ethanol sector [12–14]. Results from recent studies investigating production costs of advanced biofuels, including cellulosic ethanol, are summarized below.

These studies are: the Joint Research Centre (JRC), Eucar and Concawe (JEC) cost analysis (in publication) [37], the 2017 European Commission report by the Sub Group on Advanced Biofuels (SGAB) [38], and the 2016 report from the International Renewable Energy Agency (IRENA) [15].

For cellulosic ethanol from agricultural residues or woody biomass, cost ranges reported by the three studies (for different years) are summarized in Figure 2.

The cost ranges depend on the assumptions on the various components of the cost calculations and, in particular, on the cost of feedstock, the scale of the plants, and capital cost.

Being at an earlier stage of commercialization, cellulosic ethanol is still facing significant technology challenges and is showing significant capital and operational costs due to the complexity of the conversion processes and low maturity of the technology, but there is potential for future cost reduction [37].

The cost of feedstock is the major contributing factor depending on its accessibility, transportation costs, and also the alternative use. Feedstocks that are used in cellulosic ethanol plants have usually regional prices and they are traded locally. Their price depends on the local amount of production as well as their competing uses [39].

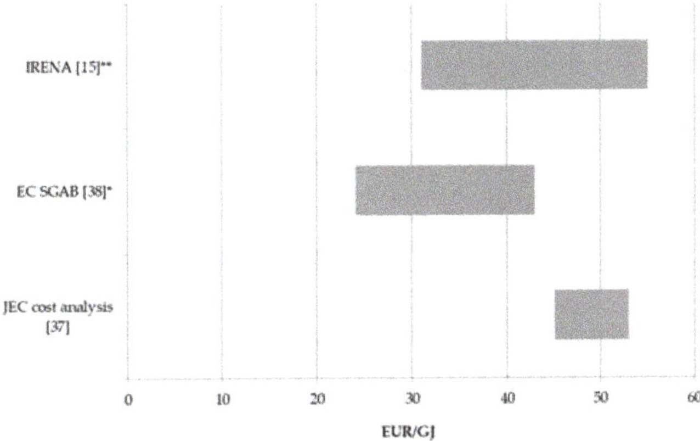

Figure 2. Production costs ranges of cellulosic ethanol from different studies. * Data extracted from Figure 18 of the report; ** Data from original source (reported for first commercial plants) has been converted from US dollar to Euro assuming 2015 exchange rate).

Cost of enzymes also represents a relevant cost component of the operational costs [15,37,40] and reducing this cost is key to making cellulosic ethanol economically viable [40]. Comparing the costs of three approaches for producing enzymes (off-site, on-site, and integrated), Johnson [40] found that the integrated method has the lowest cost.

Generally, the reported ranges of production costs for cellulosic ethanol appear to be substantially higher compared to conventional ethanol prices and far from being competitive with fossil fuel [37].

4. R&D Investment in Recent Years

This section collects information on relevant projects related to cellulosic ethanol funded under the European Union's Research and Innovation Funding programs in the last decade, namely the Seventh Framework Programme (FP7) (that covers the 2007–2013 period) and the current Horizon 2020 program (H2020) (that covers the 2014–2020 period). The projects were collected for the purpose of the Low Carbon Energy Observatory (LCEO) project, an administrative arrangement executed by DG-JRC for DG-RTD to provide data and analysis on developments in low carbon energy supply technologies. One of the technologies under analysis is the 'Sustainable advanced biofuels' sector, where cellulosic ethanol is considered. The LCEO project (started in April 2015) produces its main reports on a two-year cycle: the first set of reports was produced in 2016 (not publicly available) while the second set was prepared in 2018 and will be publicly available in 2019. In the reports, an analysis, in terms of objectives and main achievements of projects related to advanced biofuels technologies (including 'fermentation') was performed to define the project impacts on the development of the technology [41,42]. However, for the purpose of this paper, the number of considered projects has been limited exclusively to the projects which involve research on the cellulosic ethanol sector. Projects were collected from the European Commission's Community Research and Development Information Service (CORDIS) [43], that is the primary source of results from the projects funded by the EU's framework programs for research and innovation, by adopting 'cellulosic ethanol' or 'fermentation' as keywords.

Projects on cellulosic ethanol in both programs received the greatest amount of EU funding when considering all projects related to the advanced biofuel sector, with a share of around 27% in the recent H2020 program [42].

Figure 3 shows the number of FP7 (started between 2008 and 2015) and H2020 projects (started between 2015 and 2017) and the corresponding total amount of EU public funding. In general, projects in the cellulosic ethanol sector are large projects in terms of total investment needed for their implementation.

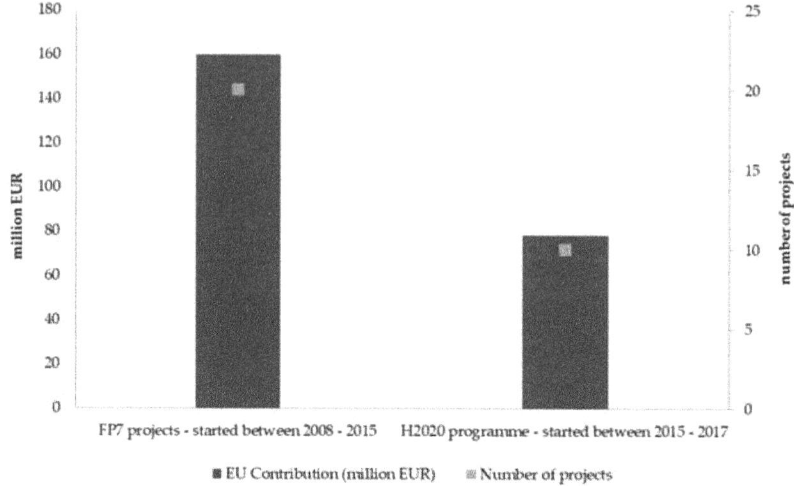

Figure 3. Numbers of Seventh Framework Programme (FP7) and Horizon 2020 program (H2020) projects and total amount of EU funding (million EUR) identified for cellulosic ethanol.

Figure 4 shows the total amount of EU funding by project coordinator countries identifying the leading countries involved in the development of this technology. The projects are mainly coordinated by Italy, Germany, and France.

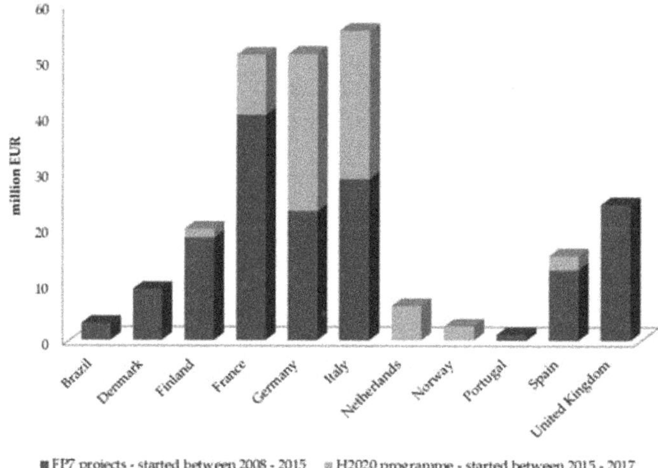

Figure 4. Total amount of EU funding by project coordinator country.

In the US, the Department of Energy (DoE) is supporting R&D projects through the Biomass Research and Development Initiative (BRDI). In 2018, DoE announced up to $3 million in funding for advanced biofuels, bioenergy, and biobased products and two selected projects received between $1 million to $2 million to develop biofuels from cellulosic ethanol and ligno-cellulosic biomass, respectively. Other funding was made available under the BioEnergy Engineering for Products Synthesis program, with a total funding of up to $28 million (in 2018), supporting projects that are aiming to create efficient conversion processes for biomass and waste derived fuels (16 projects were selected for this program). Process Development for Advanced Biofuels and Biopower is another program that supports 10 projects with $22 million. Some of the projects within this program are researching renewable fuels derived from domestic biomass feedstocks and wastes [44].

5. Technology Trends and Barriers to Large Scale Deployment

The trend in development of cellulosic ethanol production has been towards projects seeking to show (and indeed improve) upon the overall production chain; important to demonstrating the viability of this pathway. Overall, the development of both energy and cost effective pretreatment, hydrolysis and fermentation, remain the challenges hindering large-scale deployment of lignocellulosic biomass conversion to ethanol.

5.1. Pretreatment

One of the major challenges of lignocellulosic to ethanol is to ensure optimization of biomass conversion into components and by-products. Pretreatment schemes ensuring optimized use of the biomass continue to be developed. Raw material flexibility, minimum inhibitor formation, as well as maximum carbohydrate yields are central targets. Using feedstocks available in quantities large enough to supply a factory (as opposed to tiny amounts upon which a laboratory test could be carried out) brings its own technological challenges: the variability in quality and composition of the feedstock is clearly recognized as a key point, resulting in the need for a robust pretreatments section [45]. Among the various techniques applied in real plants, enzymatic hydrolysis has proven to be efficient, but enzymes cost can account for up to 30–50% of the total cost of ethanol production. In order to overcome this barrier, new types of cellulases are being studied, such as bacterial enzyme complexes. Additionally, the presence of lignin during hydrolysis behaves as an inhibitor when it deposits itself onto cellulose in biomass, making it inaccessible to enzymes, or it can even adsorb the enzymes itself [14].

5.2. Fermentation

Simultaneous utilization of pentose sugars by highly effective industrial yeast strains is still a challenge in developing continuous fermentation, which is expected to increase the yield and reduce the cost of the final product. The tolerance of ethanol producing bacteria for high substrate, inhibitor, and product concentration still needs to be improved; its progress on developments of co-fermenting microbes have been slower than projected. Review on "present technical issues" hindering cellulosic ethanol production indicated microbial species had been engineered to ferment both C5 and C6 sugars, but ethanol yields were low, and the microbes had low tolerance to high ethanol contents compared to C5 fermenting microbes [46]. Liu et al. [14] further reported that there has been considerable work trying to improve co-fermenting bacteria for at least four decades to date, but still today, it's possible to recognize familiar difficulties amongst various projects.

Other ethanol producing organisms such as yeasts, Escherichia coli, Klebsiella oxytoca, Lactobacillus sp., Clostridium sp., and others are developed for the simultaneous utilization of pentose and hexose sugars but their practical application has still to be proven. Alcohol producing strains with the ability to hydrolyze polymeric substrates have been in development for some time; although high end product concentration and selectivity (toward greater production of the desired alcohol), and insensitivity to inherent and generated inhibitors and process conditions remain major goals.

High dry matter concentration in a fermentation batch is also desirable as this will give high product concentration and help product recovery from fermentation broth. This could be achieved by developing novel process layouts involving systems aiming at immobilization of the fermenting organisms by the advanced use of non-fouling membrane systems, encapsulation of the organisms in polymer beads, etc. Efficient product separation is another advantage of advanced fermentation set-ups. This could be a critical step in the direction of a transfer from the current batch-wise into continuous fermentation processes, which would represent a more effective conversion.

5.3. Downstream

Downstream processing of products requires advances in membrane or adsorbent technology. One challenge is effective separation of alcohols from water. There is a need for membranes with high removal capacity of product, e.g., for pervaporation or suitable absorbents. Separation and rectification technology is the most demanding and needs further research on materials (membranes and adsorbents). The need to add water to the fermentation mixture to keep viscosity low and to limit bacteria inhibition creates another issue as the water will need to be driven-off during distillation. Thus, work to improve this need for water addition would likely improve the overall plant energy balance [14].

5.4. By-Products

The fate of lignin and hemicelluloses are one of the important challenges to be overcome. Processing has to avoid unfavorable conditions for sugar re-formation (back-reaction), chemical derivatization (pentoses to furfural, lignin to sulfo-lignin, and formation of lignin-carbohydrate complexes), and physical change. Separated raw material constituents (i.e., lignin and extractives) can be further converted into value-added products, for which a lot of research has been done. Fermentation broth as well as solid residues (including bacterial/yeast cell mass) are nutrient rich and can be recycled into the process, used as feed for animals, or added to biogas plants. However, it should be noted that an effective separation process for the biomass constituents after pretreatment remains a challenge.

6. Conclusions

Despite policy targets and significant R&D funding in the last decade (under EU and other programs), the cellulosic ethanol process appears to be still stagnating, mainly due to technical difficulties and high production costs that makes it uncompetitive with starch based ethanol production or fossil fuels [47]. External factors such as low oil prices may have also affected production [15]. Globally, there are several commercial scale cellulosic ethanol plants, but a substantial number of these plants are currently in idle or on-hold states.

Microbial strains which can ferment pentose and hexose sugars under large industrial-scale production are still under development, despite the past four decades of work and significant investment in R&D.

Finally, the authors note there are other possible issues potentially hindering future efficient cellulosic ethanol production, such as feedstock logistics, and storage; nevertheless, these aspects have not been analysed in this work, as they are likely to become relevant only once robust and efficient cellulosic ethanol production chains will be proven in regular operation, and at a reasonable scale.

Author Contributions: All three authors contributed significantly to the work with their expertise. M.P. led the work and performed the analysis, including the costs considerations, supported by A.C., A.C. and M.P. contributed with their expertise on technology trends and barriers.

Funding: This research received no external funding.

Acknowledgments: We wish to thank DG RTD colleagues, Maria Georgiadou and Thomas Schleker, for providing feedback on EU R&D aspects.

Conflicts of Interest: The authors declare no conflict of interest.

References

1. European Parliament and Council of the European Union. *Directive 2009/28/EC of the European Parliament and of the Council of 23 April 2009 on the Promotion of the Use of Energy from Renewable Sources and Amending and Subsequently Repealing Directives 2001/77/EC and 2003/30/EC;* European Parliament and Council of the European Union: Brussels, Belgium, 2009.

2. European Parliament and Council of the European Union. *Directive 2009/30/EC of the European Parliament and of the Council of 23 April 2009 Amending Directive 98/70/EC as Regards the Specification of Petrol, Diesel and Gas-oil and Introducing a Mechanism to Monitor and Reduce Greenhouse Gas Emissions and Amending Council Directive*

  1999/32/EC as Regards the Specification of Fuel Used by Inland Waterway Vessels and Repealing Directive 93/12/EEC;
European Parliament and Council of the European Union: Brussels, Belgium, 2009.

3. European Parliament and Council of the European Union. *Directive (EU) 2015/1513 of the European Parliament and of the Council of 9 September 2015 Amending Directive 98/70/EC Relating to the Quality of Petrol and Diesel Fuels and Amending Directive 2009/28/ EC on the Promotion of the Use of Energy from Renewable Sources*; European Parliament and Council of the European Union: Brussels, Belgium, 2015.

4. European Parliament and Council of the European Union. *Directive 2018/2001 of the European Parliament and of the Council of 11 December 2018 on the Promotion of the Use of Energy from Renewable Sources (recast)*; European Parliament and Council of the European Union: Brussels, Belgium, 2018.

5. Environmental Protection Agency (EPA). Overview for Renewable Fuel Standard. Available online: https://www.epa.gov/renewable-fuel-standard-program/overview-renewable-fuel-standard (accessed on 1 September 2019).

6. US Energy Independence and Security Act of 2007. Available online: https://www.govinfo.gov/content/pkg/PLAW-110publ140/pdf/PLAW-110publ140.pdf (accessed on 23 October 2019).

7. Ernsting, A.; Smolker, R. Dead End Road: The false promise of cellulosic biofuels. Biofuelwatch Report. 2018. Available online: https://www.biofuelwatch.org.uk/2018/dead-end-road/ (accessed on 29 August 2019).

8. US Energy Information Administration. Environmental Protection Agency (EPA) finalizes Renewable Fuel Standard for 2019, Reflecting Cellulosic Biofuel Shortfalls. 2018. Available online: https://www.eia.gov/todayinenergy/detail.php?id=37712 (accessed on 1 September 2019).

9. Barros, S.; Berk, C. *Brazil Biofuels Annual 2018*; Global Agricultural Information Network (GAIN) Report Number: BR18017; US Department of Agriculture (USDA) Foreign Agricultural Service: Washington, DC, USA, 2018.

10. The Brazilian Development Bank. BNDES and Finep Launch a R$ 1.48 bn Program to Encourage Innovation in the Sugar-Based Ethanol Sector. 2014. Available online: https://www.bndes.gov.br/SiteBNDES/bndes/bndes_en/Institucional/Press/Noticias/2014/20140217_PAISS.html (accessed on 30 June 2018).

11. Kim, G. *China Biofuels Annual 2018*; Global Agricultural Information Network (GAIN) Report Number: CH18041; US Department of Agriculture (USDA) Foreign Agricultural Service: Washington, DC, USA, 2018.

12. Badger, P.C. Ethanol from cellulose: A general review. In *Trends in New Crops and New Uses*; Janick, J., Whipkey, A., Eds.; ASHS Press: Alexandria, VA, USA, 2002; pp. 17–21.

13. Kumar, R.; Tabatabaei, M.; Karimi, K.; Sárvári Horváth, I. Recent updates on lignocellulosic biomass derived ethanol—A review. *Biofuel Res. J.* **2016**, *9*, 347–356. [CrossRef]

14. Liu, C.-G.; Xiao, Y.; Xia, X.-X.; Zhao, X.-Q.; Peng, L.; Srinophakun, P.; Bai, F.-W. Cellulosic ethanol production: Progress, challenges and strategies for solutions. *Biotechnol. Adv.* **2019**, *37*, 491–504. [CrossRef] [PubMed]

15. Alberts, G.; Ayuso, M.; Bauen, A.; Boshell, F.; Chudziak, C.; Gebauer, J.P.; German, L.; Kaltschmitt, M.; Nattrass, L.; Ripken, R.; et al. *Innovation Outlook, Advanced Liquid Biofuels*; International Renewable Energy Agency (IRENA): Abu Dhabi, UAE, 2016; Available online: https://www.irena.org/-/media/Files/IRENA/Agency/Publication/2016/IRENA_Innovation_Outlook_Advanced_Liquid_Biofuels_2016.pdf (accessed on 30 June 2018).

16. Renewable Fuels Association. Available online: http://www.ethanolrfa.org/resources/industry/statistics/#1454098996479-8715d404-e546 (accessed on 15 July 2019).

17. ePURE. European Renewable Ethanol—Key Figures. 2017. Available online: https://www.epure.org/media/1763/180905-def-data-epure-statistics-2017-designed-version.pdf (accessed on 17 July 2019).

18. Flach, B.; Lieberz, S.; Bolla, S. *EU Biofuels Annual 2019*; Global Agricultural Information Network (GAIN) Report Number: NL1902; US Department of Agriculture (USDA) Foreign Agricultural Service: Washington, DC, USA, 2019.

19. International Energy Agency (IEA) Bioenergy Task 39. Database on Facilities for the Production of Advanced Liquid and Gaseous Biofuels for Transport. Available online: https://demoplants.bioenergy2020.eu/ (accessed on 29 August 2019).

20. Schill, R.S. Zero to 10 Million in 5 Years. Ethanol Producers Magazine. 2018. Available online: http://www.ethanolproducer.com/articles/15344/zero-to-10-million-in-5-years (accessed on 16 July 2019).

21. Lane, S. Sinatra Bio: Ol' Brew Eyes Is back. Biofuels Digest. 2016. Available online: http://www.biofuelsdigest.com/bdigest/tag/synata-bio/ (accessed on 25 June 2018).

22. Sapp, M. Eni's Versalis wins Biochemtex and Beta Renewables at Auction. Biofuels Digest. 2018. Available online: http://www.biofuelsdigest.com/bdigest/2018/10/01/enis-versalis-wins-biochemtex-and-beta-ren ewables-at-auction/ (accessed on 15 July 2019).

23. Verbio Press Release. VERBIO to Acquire DuPont's Nevada, Iowa-Based Cellulosic Ethanol Plant. Press-release, 2018. Available online: https://www.verbio.de/en/press/news/press-releases/verbio-to-a cquire-duponts-nevada-iowa-based-cellulosic-ethanol-plant/ (accessed on 15 July 2019).

24. Lane, J. New Life for INEOS Bio plant: Alliance Bio-Products Wins US OK for Cellulosic Ethanol Re-fit. Biofuels Digest. 2017. Available online: http://www.biofuelsdigest.com/bdigest/2017/07/11/new-life-for-ine os-bio-plant-alliance-bio-products-wins-us-ok-for-cellulosic-ethanol-re-fit/ (accessed on 15 July 2019).

25. Schill, S.R.; Bailey, A. Inside the Cellulosic Industry, Ethanol Producer Magazine, Ethanol Producer. 2017. Available online: http://www.ethanolproducer.com/articles/14479/inside-the-cellulosic-industry (accessed on 30 June 2018).

26. Lane, J. Beta Renewables in Cellulosic Ethanol Crisis, as Group M&G Parent Files for Restructuring. Biofuel Digest. 2017. Available online: http://www.biofuelsdigest.com/bdigest/2017/10/30/beta-renewables-in-cellulosic-ethanol-crisis-as-grupo-mg-parent-files-for-restructuring/ (accessed on 30 June 2018).

27. Clariant Website. Clariant and Enviral Announce First License Agreement on Sunliquid®Cellulosic Ethanol Technology. 2017. Available online: https://www.clariant.com/en/Corporate/News/2017/09/Clariant-and-Enviral-announce-first-license-agreement-on-sunliquid-cellulosic-ethanol-technology (accessed on 15 July 2019).

28. Clariant Website. Groundbreaking for Clariant's Sunliquid®Cellulosic Ethanol Plant in Romania. Available online: https://www.clariant.com/en/Corporate/News/2018/09/Groundbreaking-for-Clariantrs quos-sunliquidreg-cellulosic-ethanol-plant-in-Romanianbsp (accessed on 15 July 2018).

29. Searle, S.; Pavlenko, N.; El Takriti, S.; Bitnere, K. *Potential Greenhouse Gas Savings from a 2030 Greenhouse Gas Reduction Target with Indirect Emissions Accounting for the European Union*; Working Paper 2017-05; International Council on Clean Transportation (ICCT): Washington, DC, USA, 2017.

30. European Parliament, Directorate-General for Internal Policies (Policy Department B: Structural and Cohesion Policies) study; Marelli, L.; Edwards, R.; Moro, A.; Kousoulidou, M.; Giuntoli, J.; Baxter, D.; Vorkapic, V.; Agostini, A.; Padella, M.; et al. *The Impacts of Biofuels on Transport and the Environment and their Connection with Agricultural Development in Europe*; European Parliament: Brussels, Belgium, 2015.

31. Valin, H.; Peters, D.; van den Berg, M.; Frank, S.; Havlik, P.; Forsell, N.; Hamelinck, C. *The Land use Change Impact of Biofuels Consumed in the EU. Quantification of Area and Greenhouse Gas Impacts, 2015*; A Cooperation of Ecofys, IIASA and E4tech. European Commission, Project number: BIENL13120; ECOFYS Netherlands: Utrecht, Netherlands, 2015.

32. California Air Resources Board (CARB), 2015. Notice of Public Hearing: Staff Report: Initial Statement of Reasons. Appendix, I. Detailed Analysis for Indirect Land use Change. 2015. Available online: https://www.arb.ca.gov/regact/2015/lcfs2015/lcfs2015.htm (accessed on 1 September 2019).

33. US Environmental Protection Agency (EPA). 40 CFR Part 80. Regulation of Fuels and Fuel Additives: Changes to Renewable Fuel Standard Program; Final Rule. 2010. Available online: https://www.gpo.gov/fd sys/pkg/FR-2010-03-26/pdf/2010-3851.pdf (accessed on 1 September 2019).

34. Laborde, D. *Assessing the Land use Change Consequences of European Biofuel Policy*; International Food Policy Research Institute: Washington, DC, USA, 2011.

35. Searle, S. *Sustainability Challenges of Lignocellulosic Bioenergy Crops*; International Council on Clean Transportation (ICCT): Washington, DC, USA, 2018.

36. Pavlenko, N.; Searle, S. *A Comparison of Induced Land use Change Emissions Estimates from Energy Crops*; International Council on Clean Transportation (ICCT): Washington, DC, USA, 2018.

37. Padella, M.; Edwards, R.; Candela Ripoll, I.; Lonza, L. *Estimates of Biofuel Production Costs and Cost of Savings—Annex of JEC Well-to-Tank Version 5*, in publication.

38. European Commission, Sub Group on Advanced Biofuels (SGAB). *Sustainable Transport Forum. Building Up the Future, Final Report*; European Commission: Brussels, Belgium, 2017.

39. Chudziak, C.; Alberts, G.; Bauen, A. Ramp Up of Lignocellulosic Ethanol in Europe to 2030. In Proceedings of the E4Tech report for the Co-Sponsors of the 6th International Conference on Lignocellulosic Ethanol (BetaRenewables DuPont, ePURE, Leaf, Novozymes, Shell St1), Brussels, Belgium, 27–28 September 2017.

40. Johnson, E. Integrated enzyme production lowers the cost of cellulosic ethanol. *Biofuels. Bioprod. Bioref.* **2016**, *10*, 164–174. [CrossRef]

41. Rocca, S.; Padella, M.; O'Connell, A.; Giuntoli, J.; Kousoulidou, M.; Baxter, D.; Marelli, L. Technology Development Report Sustainable Advanced Biofuels 2016, Deliverable D2.1.12 for the Low Carbon Energy Observatory. *JRC Sci. Policy Rep.* **2016**. (internal report).

42. Padella, M.; O'Connell, A.; Prussi, M.; Flitris, E.; Lonza, L. Technology Development Report Sustainable Advanced Biofuels 2018, Deliverable D2.2.12 for the Low Carbon Energy Observatory. in publication.

43. CORDIS Website. Available online: https://cordis.europa.eu/ (accessed on 15 February 2018).

44. US Office of Energy Efficiency & Renewable Energy. Available online: https://www.energy.gov/eere/bioenergy/bioenergy-technologies-office-closed-funding-opportunities#2019_1 (accessed on 16 July 2019).

45. Chiaramonti, D.; Prussi, M.; Ferrero, S.; Oriani, L.; Ottonello, P.; Torre, P.; Cherchi, F. Review of pretreatment processes for lignocellulosic ethanol production, and development of an innovative method. *Biomass Bioenergy* **2012**, *46*, 25–35. [CrossRef]

46. Yusuf, F.; Gaur, N.A. Engineering Saccharomyces cerevisiae for C5 fermentation: A step towards second-generation biofuel production. In *Metabolic Engineering for Bioactive Compounds*; Springer: Singapore, 2017; pp. 157–172.

47. Rapier, R. Cellulosic Ethanol Falling Far Short of the Hype, R-Squared Energy. Available online: http://www.rrapier.com/2018/02/cellulosic-ethanol-falling-far-short-of-the-hype/ (accessed on 25 June 2018).

Article

Analysis of Internal Gas Leaks in an MCFC System Package for an LNG-Fueled Ship

Gilltae Roh [1,2], Youngseung Na [3], Jun-Young Park [4] and Hansung Kim [2,*]

[1]   Future Technology Research Team, Korean Register (KR), Busan 46762, Korea; gtroh@krs.co.kr
[2]   Department of Chemical and Biomolecular Engineering, Yonsei University 134 Shinchon-Dong, Seodaemun-gu, Seoul 120-749, Korea
[3]   Department of Mechanical and Information Engineering, University of Seoul, Seoulsiripdaero 163, Dongdaemun-gu, Seoul 02504, Korea; ysna@uos.ac.kr
[4]   HMC, Department of Nanotechnology and Advanced Materials Engineering, Sejong University, Seoul 143-747, Korea; jyoung@sejong.edu
*   Correspondence: elchem@yonsei.ac.kr; Tel.: +82-2-2123-5753; Fax: +82-2-312-6401

Received: 19 April 2019; Accepted: 3 June 2019; Published: 6 June 2019

Abstract: The airflow inside the housing of a 300-kW molten carbonate fuel cell (MCFC) system is designed to ensure safety in case of a gas leak by applying computational fluid dynamics (CFD) techniques. In particular, gas accumulating zones are identified to prevent damage to vulnerable components from high temperature and pressure. Furthermore, the location of the alarm unit with the gas-leak detector is recommended for construction of safe MCFC ships. In order to achieve this, a flow-tracking and contour field (for gas, temperature, and pressure) including a fuel-cell stack module, balance-of-plant, and various pipes is developed. With the simulated flow field, temperature flow is interpreted for the heating conditions of each component or pipe in order to find out where the temperature is concentrated inside the fuel cell system, as well as the increase in temperature at the exit. In addition, the gas leakage from the valves is investigated by using a flow simulation to analyze the gas and pressure distribution inside the fuel cell system.

Keywords: ship structure; LNG-fueled ship; green ship; numerical analysis; flow characteristics; molten carbonate fuel cell system

---

1. Introduction

The policies for marine environmental regulations of the International Maritime Organization (IMO) and governments have recently been strengthened to reduce air contaminants and greenhouse gases, including nitrogen oxides (NO_x), sulfuric oxides (SO_x), and carbon dioxide (CO_2). Marine engineers have therefore considered various strategies for building green ships such as developing a highly efficient propeller, modification of fan shapes, and optimization of an operational window [1–3]. In particular, in 2016, the IMO Marine Environment Protection Committee (MEPC) restricted the emissions of SO_x from 3.5% to 0.5% for marine ships in the entire ocean until 2020 [4]. Today, many shipping firms are adopting liquefied natural gas (LNG) as marine fuel in an attempt to replace conventionally used heavy fuel oil (HFO). The use of LNG in ships is considered to be a way to safely use boil-off gas (BOG) that is naturally generated by heat leaks. As one option, the application of fuel cells to ships is being considered [5]. However, the ultimate goal of shipbuilders is to replace diesel engines, which emit greenhouse gases, with regenerative energy generators (zero-emission ships).

Energy storage systems (ESSs), and sunlight have received significant attention as environmentally friendly energy sources for green ships in the future [6]. Among these, fuel cells (FCs) have strong potential as an alternative to traditional marine propulsion power plants owing to their high efficiency, easy modularization, multi-fuel flexibility, and environmental friendliness [7]. In addition, FCs can essentially offer a silent and vibration-free operation, reducing the need for noise insulation of machinery. In particular, high-temperature molten carbonate fuel cells (MCFCs) have various advantages for green ships, because they achieve a system efficiency of more than 80%, with negligible air-pollutant emissions when the waste heat is recovered. Furthermore, it may be economically efficient to use fuels such as LNG and liquefied hydrogen (H_2) carriers in gas-fueled and gas carriers, because MCFCs can use the BOG emitted from the LNG ship [8,9]. However, the development of green ships using fuel cells is still at a primitive stage, because of the considerable cost and time needed for construction of ships. Furthermore, liquefied gas carriers and gas-fueled ships may pose a high risk of explosion and fire from gas leaks, resulting from the localized damage (thermal expansion) of system components caused by high heat, because MCFCs are operated at a high temperature [10]. In addition, in order to use fuel cells in gas-fueled and gas carriers, severe weather and sea winds are considered among the design parameters needed to build a highly stable green ship with fuel cells.

In this regard, mathematical modeling plays an important role by significantly reducing the need for repetitive experiments and confirming concepts for building ships [11–13]. Moreover, several technical issues associated with the high operating temperature, such as explosion and fire caused by high heat, can be explored and thus mitigated through the new design and development progress for MCFC green ships. Ovrum and Dimopoulos (2012) [14] first reported the modeling work for an MCFC auxiliary power unit installed on board a vessel. They developed a modular dynamic model consisting of a set of partial differential and non-linear algebraic equations, employing 2D and 3D solid geometry and real gas properties to provide insight into the process dynamics and cell performance. Marra et al. [15] also presented the modeling approach as a tool for improving the current collector and gas distributor designs, optimizing MCFC performance with new technological solutions. This approach played a fundamental role in the thermal management of the MCFC, which is one of the most crucial points in performance improvements. However, there have been no reports of a safety study on the application of the MCFC system to a ship that comprehensively evaluates a computational fluid dynamics (CFD) analysis while considering intentional leakage simulation in the marine environment. Analysis of intentional leakage can indicate the design guidelines for a safe system and thus prevent potential damage.

Hence, to design highly efficient and stable MCFC green ships, we analyzed the characteristics (flows and distribution) of gas, heat, and pressure for the FC systems, e.g., stack, pipes, heat exchangers (HEXs) and balance-of-plant (BOP), by using the simulation tools referred to in the open literature. In particular, zones of localized high gas concentration, temperature, and pressure are identified, and design parameters needed to build safe MCFC ships, such as the location of the alarm unit with the gas leak detector and the capacity of exhaust fans, are recommended to prevent damage to system components from high heat and pressure. This study is expected to significantly contribute to the increased practical use of MCFCs in merchant marine applications.

Figure 1 is a schematic diagram of a fuel-cell system (BOP and stack) as an auxiliary power unit that can produce 300 kW of electricity by using the boil-off gas (BOG) of the LNG carrier. The BOG (as the fuel) generated from the LNG tank is fed to the pre-converter and anode side, and air (oxidant) is supplied from the cathode side to produce 300 kW of electricity through chemical and electrochemical reactions in the stack. The reactions are as follows:

$$\text{Anode}: H_2 + CO_3^{2-} \leftrightarrow H_2O + CO_2 + 2e^- \tag{1}$$

$$CO + CO_3^{2-} \leftrightarrow 2CO_2 + 2e^- \tag{2}$$

$$CH_4 + H_2O \leftrightarrow CO + 3H_2 \tag{3}$$

$$CO + H_2O \leftrightarrow CO_2 + H_2 \tag{4}$$

$$\text{Cathode}: CO_2 + \frac{1}{2}O^2 + 2e^- \leftrightarrow CO_3^{2-} \tag{5}$$

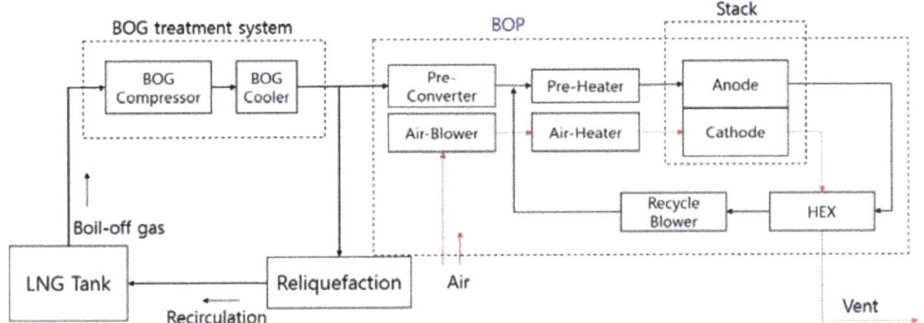

Figure 1. Scheme of boil-off gas (BOG) treatment system with fuel cell system for a liquefied natural gas (LNG)-fueled ship.

In this study, the characteristics of flow inside the 300-kW fuel-cell system was analyzed using CFD. To prevent gas leaks from inside the BOP, it is very important to identify the status of internal air flow and to find an appropriate location for installing the equipment for gas detection.

2. CFD Simulation Method

2.1. Modeling and Boundary Conditions

Figure 2 shows a 3D CAD model of the components inside the BOP, which is designed to operate the 300-kW MCFC stack. Inside the BOP is a pretreatment device used to provide fuel to the anode and oxidant to the cathode. The anode side of the BOP includes the pre-converter, polisher, humidifier, recycle blower, etc. The pre-converter converts the methane from the BOG to about 10% hydrogen to supply to the stack. The humidifier produces steam, which is supplied to the anode side of the stack for the electrochemical reaction in the stack. The polisher and recycle blower recirculate the exhaust gas at the end of the stack.

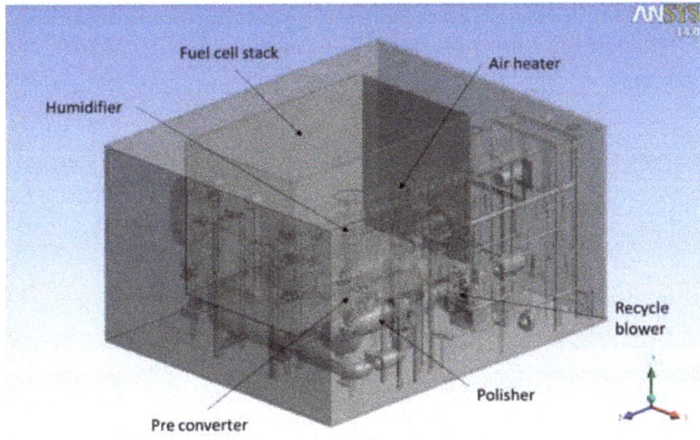

Figure 2. Component locations in a molten carbonate fuel cell (MCFC) stack package.

The grid system of the package (without the external components) was created using the ANSYS Design Modeler [16] and ICEM-CFD [17]. An unstructured tetrahedron grid technique was efficiently applied to create the overall grid system. The grid cells with a prism shape were applied near a specific solid wall to resolve the relatively high velocity gradient. The uniform thickness ratios were maintained for the grid cell layers with different unit cells. Thus, in this analysis, approximately 20 million grid cells were used. The solution was independent from computational cell numbers.

The louvers on the outer walls of the package room shown in Figure 3a were installed to assist in supplying air to the inside of the room to cool the equipment. The blue parts of the figure indicate the points where air is supplied to the room. The room air was set to atmospheric pressure and an ambient temperature of 25 °C. As shown in Figure 3b, the air heated by the devices inside the package was released through the exhaust fans installed on the roof. As shown in Figure 3b, in order to release the air at a speed of 10.3 m·s^{-1} through each exhaust pipe, the performance of the applied exhaust fan was 5000 m^3·h^{-1} and the cross-sectional area of the exhaust pipes was 0.135 m^2. The heating surface (shown in blue in Figure 2) was set according to the specific heat fluxes for the heating area shown in Table 1. Since the package room is exposed to the outside, heat is transferred through the walls according to the surroundings; and the heat-transfer coefficient was set as 100 W·m^{-2}K^{-1} to simulate the convective heat transfer of a weak wind. To identify the leakage in the system, fuel was modeled as leaked intentionally at the red points in Figure 4 with listed gas composition in Table 2. The leaked gas path line is traced by the CFD, and provides the supporting data to locate gas leak detectors.

(a) **(b)**

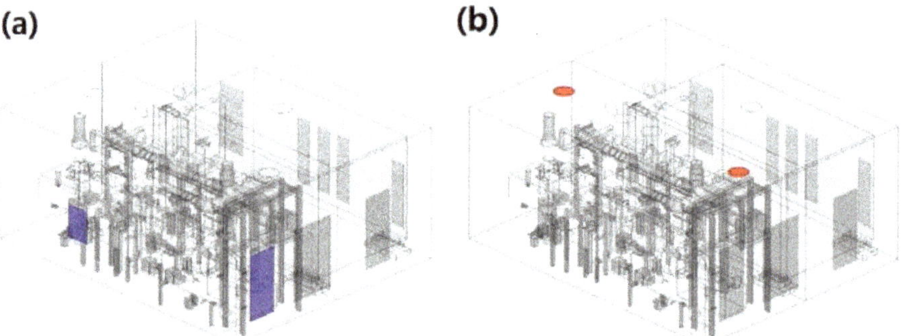

Figure 3. Conditions of the inlet and outlet of the balance-of-plant (BOP) package: (**a**) louver installation locations (inlet) and (**b**) exhaust pipe locations (outlet).

Table 1. Heat flux of the heating area of the stack package model.

	Area [m^2]	Total Heating Area [m^2]	Heat Value [kW]	Heat Flux [W/m^2]
Humidifier	1.189	1.515	1.00	660.1
EG-2012	0.326			
Polisher	1.134	2.651	2.426	915.1
EG-2011	1.517			
EG-2010	7.281	7.281	4.00	549.3
Pre-converter	2.507	8.368	2.00	239.0
FG-2008	1.959			
FG-2009	3.902			
Recycle blower	1.865	9.537	1.68	176.2
EG-4008	1.484			
EG-4009	5.689			
Air heater	0.499			
Stack module	53.111	53.111	9.00	169.5

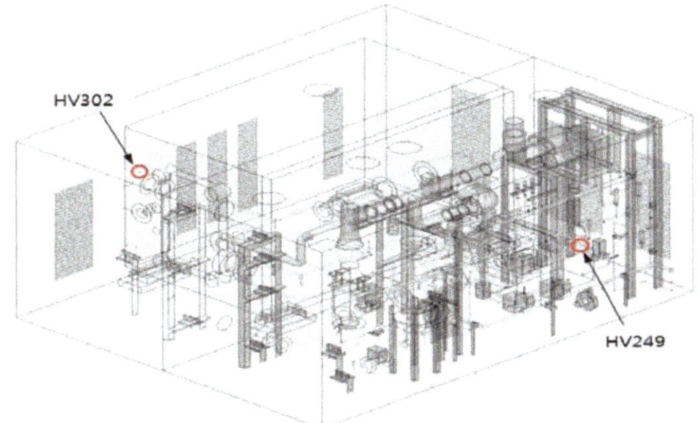

Figure 4. Leaked valve positions in the MCFC stack package.

Table 2. Leaked gas composition.

	HV249		HV302	
	Mole Fraction	**Mass Fraction**	**Mole Fraction**	**Mass Fraction**
CH$_4$	0.938	0.8934	0.1498	0.1751
H$_2$O	1.134	2.651	0.3196	0.4196
H$_2$	-	-	0.4127	0.0606
CO$_2$	-	-	0.0874	0.2803
Air	0.062	0.1066	0.0305	0.0644

2.2. Governing Equations

The governing equations of mass, momentum, turbulence, and heat transfer were applied in accordance with the buoyancy effect in the internal airflow simulation. In Reynolds averaging, the velocity vector **v** is decomposed into the mean and fluctuating velocity components as $\mathbf{v} = \bar{\mathbf{v}} + \mathbf{v}\prime$. Likewise, a scalar, ϕ (such as pressure, energy, and other quantities) is decomposed as $\phi = \bar{\phi} + \phi'$. The incompressible Reynolds-averaged Navier–Stokes (RANS) equations are defined as follows [18]:

$$\frac{\partial \rho}{\partial t} + \nabla \cdot (\rho \bar{\mathbf{v}}) = 0 \tag{6}$$

$$\frac{\partial}{\partial t}(\rho \bar{\mathbf{v}}) + \nabla \cdot (\rho \bar{\mathbf{v}}\bar{\mathbf{v}}) = -\nabla \bar{p} + \nabla \cdot (2\mu \bar{\mathbf{s}} - \rho \overline{\mathbf{v}\prime\mathbf{v}\prime}) + \rho \mathbf{g} \tag{7}$$

where ρ is the density calculated from the ideal gas equation of state, t is time, $\bar{p}$ is the mean pressure, μ is the molecular viscosity and $\bar{\mathbf{s}} = \frac{1}{2}\left(\nabla \bar{\mathbf{v}} + \nabla \bar{\mathbf{v}}^T\right)$ is the mean strain-rate tensor. The quantity $-\rho \overline{\mathbf{v}\prime\mathbf{v}\prime}$ is defined as the Reynolds stress tensor, and $\rho \mathbf{g}$ is the gravitational body force. To close the RANS equations, the Reynolds stress tensor is modeled on the basis of the Boussinesq approximation as follows:

$$\rho \overline{\mathbf{v}\prime\mathbf{v}\prime} = 2\mu_t \bar{\mathbf{s}} - \frac{2}{3}\rho K \mathbf{I} \tag{8}$$

where μ_t is the turbulent viscosity, the turbulent kinetic energy is defined as $K = \frac{1}{2}\overline{\mathbf{v}\prime\mathbf{v}\prime}$, and **I** is the unit tensor.

Turbulent kinetic energy K and turbulence dissipation rate ε for the turbulence model were calculated using Equations (9) and (10) [18].

$$\frac{\partial(\rho K)}{\partial t} + \nabla \cdot (\rho \mathbf{v} K) = \nabla \cdot \left[\left(\mu + \frac{\mu_t}{\mathrm{Pr}_k}\right)\nabla K\right] + P_k - \rho\varepsilon \tag{9}$$

$$\frac{\partial(\rho\varepsilon)}{\partial t} + \nabla \cdot (\rho \mathbf{v}\varepsilon) = \nabla \cdot \left[\left(\mu + \frac{\mu_t}{\mathrm{Pr}_\varepsilon}\right)\nabla\varepsilon\right] + \frac{\varepsilon}{k}(C_{\varepsilon 1}P_k - C_{\varepsilon 2}\rho\varepsilon) \tag{10}$$

where P_k is the generation of turbulent kinetic energy caused by the mean velocity gradients and buoyancy, $C_{\varepsilon 1}$ and $C_{\varepsilon 2}$ are the model constants, and Pr_k and Pr_ε are the turbulent Prandtl numbers for K and ε, respectively. Here, the turbulent viscosity μ_t is calculated by combining K and ε as $\mu_t = \rho C_\mu \frac{K^2}{\varepsilon}$, where C_μ is a constant.

The energy equation is modeled using the concept of the Reynolds analogy to turbulent momentum transfer:

$$\frac{\partial}{\partial t}\left(\rho C_p \overline{T}\right) + \nabla \cdot \left(\rho C_p \overline{\mathbf{v} T}\right) = \nabla \cdot \left[k\nabla T - \rho C_p \overline{\mathbf{v}' T'}\right] + S_T \tag{11}$$

where C_p is the heat capacity, T is the temperature, and k is the molecular thermal conductivity. The turbulent thermal conductivity is calculated as $k_t = \mu_t \frac{C_p}{\mathrm{Pr}_t}$, in which Pr_t is the turbulent Prandtl number. The $\rho C_p \overline{\mathbf{v}' T'}$ term is defined as $k_t \nabla T$ and the source term S_T includes a volumetric heat source.

3. Results and Discussion

3.1. Concentration Distribution

Methane was assumed to be leaked from the HV249 and HV302 valves. The gas from HV249 diffused throughout the entire chamber. The volume fraction of methane was plotted from 0 to 5%, which is the lower limit to explode, as shown in Figure 5a. Near HV302, methane was located at the point because of the small composition in the leaked gas. The iso-surface of the methane volume fraction of 2% covered around one quarter of the BOP chamber, which should be ventilated by carrier gas from the louver on the right-hand side in Figure 5b. Therefore, the gas detector should be located on the ceiling, and the evacuating jet fan should operate in an emergency to suck flammable gas from the chamber.

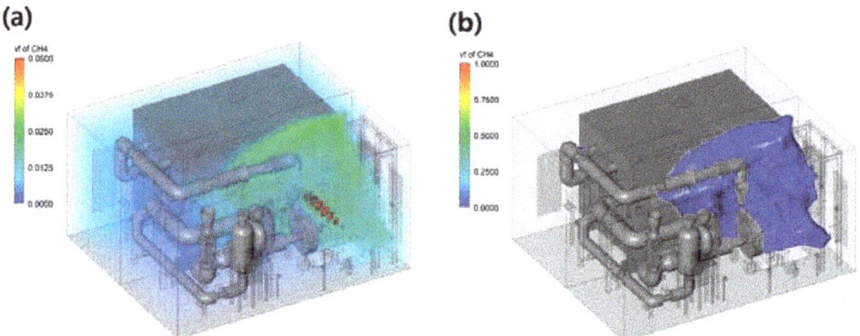

Figure 5. (a) Volume fraction distribution and (b) 2% iso-surface of leaked methane inside the MCFC stack package.

In contrast, hydrogen leakage from HV302 differed from the methane distribution. The average hydrogen concentration was much lower than that of methane and it quickly diffused to the whole chamber, as shown in Figure 6a,b. The low concentration of 1% which accumulated behind the stack because of stagnant flow is not dangerous, but it must be circulated by a blowing or suction fan.

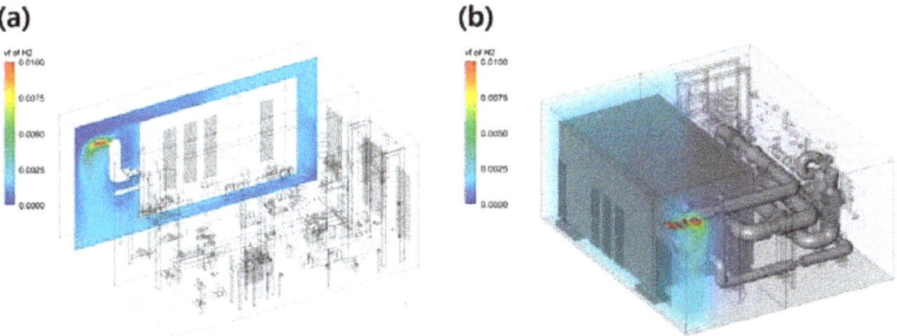

Figure 6. (**a**) One vertical section and (**b**) horizontal sections of the volume fraction of leaked hydrogen inside the MCFC stack package.

3.2. Flow Fields

Figure 7 shows the velocity vectors of the vertical sections of inlet and outlet flow. Inlet gas from the louvers on the right-hand side flowed to the holes on the ceiling, as shown in Figure 7. The average speed from the louvers was 4.2 m·s^{-1}, whereas the whole range covered from zero to 5 m·s^{-1}. The red spot in the chamber observed near HV249 indicates leaked jet flow. The left-hand side louvers on the front panel blew at an average speed of 3.8 m·s^{-1}, which was lower than the right-hand side because of the structure. The wing side of the louvers was positioned at 45° towards the internal side of the package, and the path lines formed in this direction passed through the louvers. The colors of the path lines in the figure represent the temperature distribution, showing that the temperature of the cables passing through the internal components had increased.

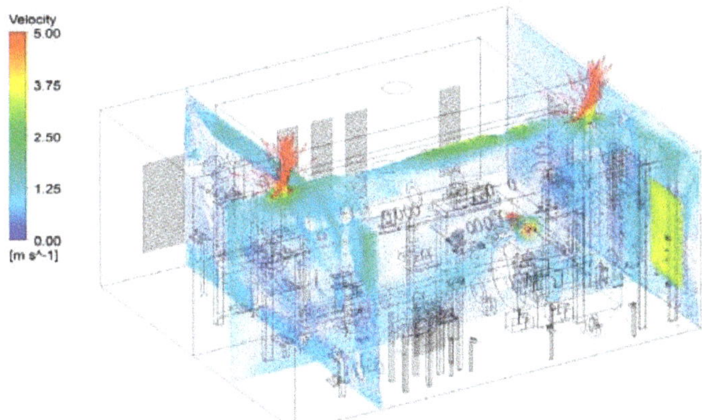

Figure 7. Velocity vectors of the MCFC stack package.

Path lines from the right-hand side louvers scattered rigorously to the outside and carried methane from the chamber in Figure 8a,b. In contrast, the left-hand side flow field has a few path lines with high temperature, because of the stagnant flow near the stack. Even though the concentration of hydrogen is less than the lower limit for an explosion, circulating flow is necessary to flush out the flammable gas from the chamber.

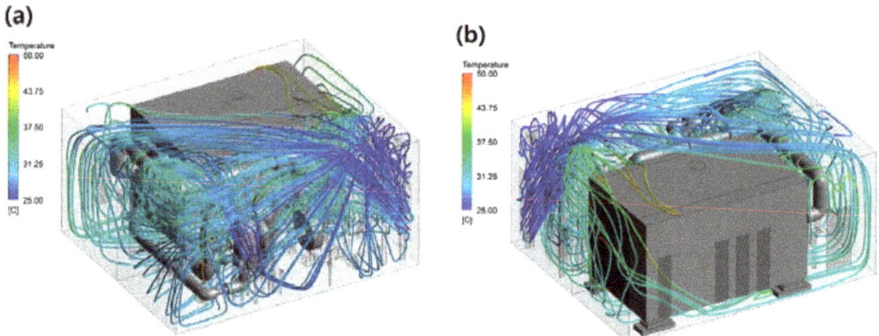

Figure 8. Path lines of inlet flows from louvers and outlet flows to the exit holes on the ceiling (**a**) frontal view and (**b**) backside view.

3.3. Temperature Fields

The skin temperature of the polisher and humidifier components, from the ambient temperature to the highest temperature of 200 °C, is shown in Figure 9a,b and indicates that the air temperature distribution in the vertical sections of the chamber was not uniform, despite the cooling air from the louvers. Because the BOP package is in contact with internal heat, the air temperature in the upper part of the package is a result of the influence of the increased density and upward flow of the exhaust air. In particular, the temperature of the upper side of the fuel cell stack was as high as that of the polisher because of accumulating flow, which relates to the flow field. Additionally, the hot leaked flow from the HV302 was stagnant near the stack. The average outlet temperature of the two exit holes was about 31.5 °C.

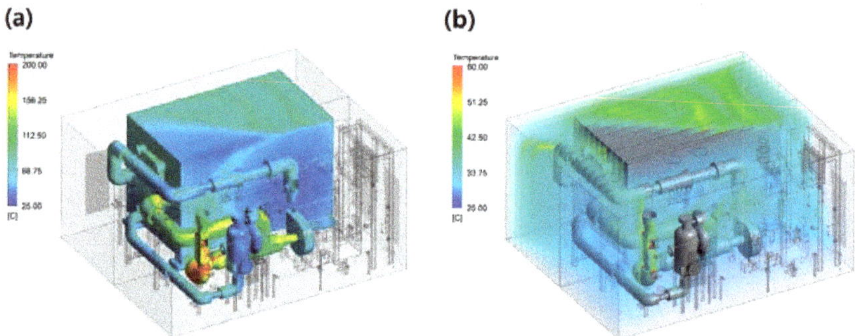

Figure 9. Temperature distribution of the MCFC stack package at outer temperature 25 °C: (**a**) surface temperature at heating pipe and (**b**) yz plane.

3.4. Pressure Fields

Figure 10 shows the vertical section of pressure distribution inside the package. The range of pressure was shown to be 101,313–101,320 Pa, and it was observed that low pressure was caused by the velocity increment near the exhaust pipe region, which lead to a static pressure drop. However, it can be concluded that the pressure differences inside the package were relatively insignificant.

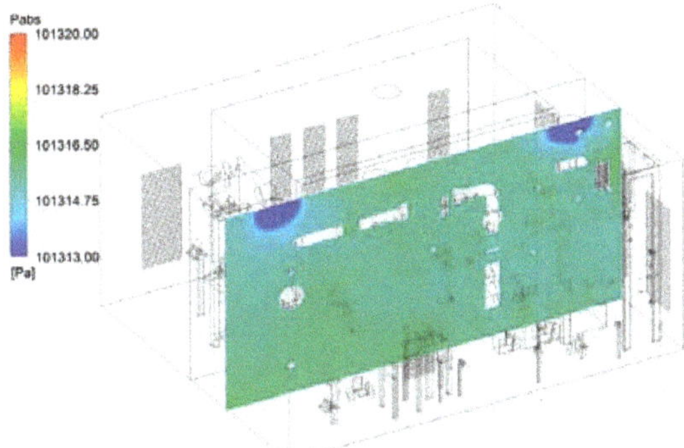

Figure 10. Pressure distribution at vertical section inside the MCFC stack package at 25 °C.

4. Conclusions

The flow characteristics inside a fuel cell balance-of-plant (BOP) package were analyzed through computational fluid dynamics (CFD) techniques. The simulation was used to interpret the characteristics of gas, pressure, and temperature flows according to chemical species. The developed model indicated that the inside flow structure plays an important role not only in heating the internal units, but also in the distribution and emission of the leaked gas. In addition, the heat flow was confirmed to be controlled, excluding the entrance and exit of the stack module. However, the flow field through the stack module, the outer walls of the package, and the top of the module were not cooled to the ambient temperature. Moreover, temperature increases were observed at the rear side of the stack module, in the stagnant area at the top of the stack module, and within the leaked gas accumulated in these areas. Therefore, the density distribution can be decreased by lowering the temperature inside the package and quickly venting the leaked gas. The capacity to supply external air should thus be designed by generating a smooth flow between the stack module and the outer walls of the package.

The interpretation of the 300-kW MCFC BOP package model shows that the flow of air generated inside the package and the surface temperature of the heating apparatus influenced the flow of heat and air in the package. In other words, if the heat generated is higher than that of the exothermic device, the temperature of the surface of the heated ventilation system would need to be kept at less than the ambient temperature. In this case, the design of the airflow inside the device is important for preventing an increase in the temperature of the internal device.

The simulation of gas leakage and fuel gas density in the BOP package has indicated that the natural gas line should be ventilated to the outside of the chamber because methane diffused in all directions throughout the whole chamber, whereas the hydrogen gas flowed directly upwards. Therefore, the safety ventilation outlet should be located with the gas alarm on the ceiling near the hydrogen pipeline. For maximum safety, the ventilation and alarm systems should be designed according to the simulated flow field.

Author Contributions: G.R. wrote the manuscript, conceptualized, and analyzed; Y.N. and J.-Y.P. analyzed and validated results; and H.K. supervised.

Funding: This material is based upon work supported by the Ministry of Trade, Industry & Energy (MOTIE, KOREA) under industrial Core technology Development Program. NO. 10070159, 'Development of CCS technologies (design and verification) to apply LH$_2$ Carrier.'

Conflicts of Interest: The authors declare no conflict of interest.

References

1. Boswinkel, H.H. *Ship Propulsion with Fuel Cells, ECN-C-91-022*; ECN: Rotterdam, The Netherlands, 1991.
2. Buhaug, Ø.; Corbett, J.; Endresen, Ø.; Eyring, V.; Faber, J.; Hanayama, S.; Lee, D.; Lee, D.; Lindstad, H.; Markowska, A. *Second IMO GHG Study 2009*; International Maritime Organization (IMO): London, UK, 2009; p. 20.
3. Sattler, G. Fuel cells going on-board. *J. Power Sources* **2000**, *86*, 61–67. [CrossRef]
4. The 70th Session of the Marine Environment Protection Committee. Amendments to the annex of the protocol of 1997 to amend the international convention of plooution from ships, 1973, as modified by the protocol of 1978 relating thereto. In *Proceedings "Resolution MEPC.278(70)"*; IMO: London, UK, 2016.
5. De-Troya, J.J.; Álvarez, C.; Fernández-Garrido, C.; Carral, L. Analysing the possibilities of using fuel cells in ships. *Int. J. Hydrog. Energy* **2016**, *41*, 2853–2866. [CrossRef]
6. Lan, H.; Bai, Y.; Wen, S.; Yu, D.; Hong, Y.Y.; Dai, J.; Cheng, P. Modeling and stability analysis of hybrid PV/diesel/ESS in ship power system. *Inventions* **2016**, *1*, 5. [CrossRef]
7. Sundmacher, K. *Molten Carbonate Fuel Cells: Modeling, Analysis, Simulation, and Control*; Wiley-VCH: Weinheim, Germany, 2007; pp. 519–932.
8. Lee, J.H.; Kim, Y.J.; Hwang, S. Computational study of LNG evaporation and heat diffusion through a LNG cargo tank membrane. *Ocean Eng.* **2015**, *106*, 77–86. [CrossRef]
9. Roh, S.; Son, G.; Song, G.; Bae, J. Numerical study of transient natural convection in a pressurized LNG storage tank. *Appl. Therm. Eng.* **2013**, *52*, 209–220. [CrossRef]
10. Ling-Chin, J.; Roskilly, A.P. Investigating the implications of a new-build hybrid power system for Roll-on/Roll-off cargo ships from a sustainability perspective—A life cycle assessment case study. *Appl. Energy* **2016**, *181*, 416–434. [CrossRef]
11. Lataire, E.; Vantorre, M.; Delefortrie, G.; Candries, M. Mathematical modelling of forces acting on ships during lightering operations. *Ocean Eng.* **2012**, *55*, 101–115. [CrossRef]
12. Wang, X.G.; Zou, Z.J.; Hou, X.R.; Feng, X. System identification modelling of ship manoeuvring motion based on support vector regression. *J. Hydrodyn. Ser. B* **2015**, *27*, 502–512. [CrossRef]
13. Welaya, Y.M.; El Gohary, M.M.; Ammar, N.R. A comparison between fuel cells and other alternatives for marine electric power generation. *Int. J. Nav. Archit. Ocean Eng.* **2011**, *3*, 141–149. [CrossRef]
14. Ovrum, E.; Dimopoulos, G. A validated dynamic model of the first marine molten carbonate fuel cell. *Appl. Therm. Eng.* **2012**, *35*, 15–28. [CrossRef]
15. Wang, K.; Hissel, D.; Péra, M.; Steiner, N.; Marra, D.; Sorrentino, M.; Pianese, C.; Monteverde, M.; Cardone, P.; Saarinen, J. A review on solid oxide fuel cell models. *Int. J. Hydrog. Energy* **2011**, *36*, 7212–7228. [CrossRef]
16. Atmaca, M.; Girgin, İ.; Ezgi, C. CFD modeling of a diesel evaporator used in fuel cell systems. *Int. J. Hydrog. Energy* **2016**, *41*, 6004–6012. [CrossRef]
17. Fluent, A. *12.0 Theory Guide*; Ansys Inc.: Pittsburgh, PA, USA, 2009; Volume 5.
18. Wilcox, D.C. *Turbulence Modeling for CFD*, 2nd ed.; DCW Industries: La Cañada Flintridge, CA, USA, 2004; pp. 35–36, 123–125.

Article

Analysis of a Supercapacitor/Battery Hybrid Power System for a Bulk Carrier

Kyunghwa Kim [1,*], **Juwan An** [1,2], **Kido Park** [1,2], **Gilltae Roh** [1] and **Kangwoo Chun** [1,*]

1   Future Technology Research Team, Korean Register (KR), Busan 46762, Korea; juwan@krs.co.kr (J.A.); kdpark@krs.co.kr (K.P.); gtroh@krs.co.kr (G.R.)
2   Division of Marine System Engineering, Korea Maritime and Ocean University, Busan 49112, Korea
*   Correspondence: kimkh@krs.co.kr (K.K.); kwchun@krs.co.kr (K.C.)

Received: 4 March 2019; Accepted: 10 April 2019; Published: 14 April 2019

Abstract: Concerns about harmful exhaust emissions from ships have been an issue. Specifically, the emissions at ports are the most serious. This paper introduces a hybrid power system that combines conventional diesel generators with two different energy storage systems (ESSs) (lithium-ion batteries (LIB) and supercapacitors (SC)) focused on port operations of ships. To verify the proposed system, a bulk carrier with four deck cranes is selected as a target ship, and each size (capacity) of LIB and SC is determined based on assumed power demands. The determined sizes are proven to be sufficient for a target ship through simulation results. Lastly, the proposed system is compared to a conventional one in terms of the environmental and economic aspects. The results show that the proposed system can reduce emissions (CO_2, SO_X, and NOx) substantially and has a short payback period, particularly for ships that have a long cargo handling time or visit many ports with a short-term sailing time. Therefore, the proposed system could be an eco-friendly and economical solution for bulk carriers for emission problems at ports.

Keywords: hybrid power system; lithium-ion battery (LIB); supercapacitor (SC); alternative maritime power (AMP); bulk carrier

1. Introduction

Although road vehicles currently represent about 70% of total greenhouse gas (GHG) emissions in the transport sector, other forms of transport—including aviation, maritime, and off-road vehicles—are also substantial emissions sources and are expected to see continued growth in the coming years. Specifically, the maritime sector is expected to rise gradually because of its slower improvement efficiency compared to other vehicles; its GHG share is expected to increase from 10% in 2018 to 20% in 2060 among the global transport-related GHG emissions [1].

In this regard, many countries and the international maritime organization (IMO) have been implementing environmental regulations or policies, especially in emission-control areas (ECAs), which are designated areas near ports where ships are required to further reduce emissions. For example, the sulfur limit is currently 0.1% within ECAs; it is 35 times stricter than the outside ECAs. These strict regulations are related to the fact that premature deaths have been increasing each year due to cardiopulmonary disease and lung cancer caused by pollutants emitted from ships at ports [2]. Shipping emissions in East Asia accounted for 16% of global shipping CO_2 in 2013, compared to only 4%–7% in 2002–2005. This increase in emissions resulted in large adverse health impacts, with 14,500–37,500 premature deaths per year [3].

Therefore, ship owners have been making efforts to reduce harmful emissions using exhaust gas treatment systems such as sulfur dioxide (SO_X) scrubbers, selective catalyst reactors (SCRs), or changing ship fuels from heavy fuel oil (HFO) to liquefied natural gas (LNG) or marine gas oil (MGO), etc. In addition, major ports have been expanding shore power facilities (or alternative maritime power

(AMP)), which can supply electric power for ships from land-based electric power plants while staying at a port. Notably, low voltage AMP facilities have already been installed in many dominant ports worldwide. Additionally, high voltage (3.3kV, 6.6kV, 11kV, etc.) AMP facilities are being installed in major ports for large ships such as in the U.S., Canada, European countries, China, etc., and the European Union (EU) requires European ports to offer shore-based electricity to ships by 2025.

In this regard, hybrid systems using an energy storage system (ESS) have gained attention as an alternative solution to solve the environmental issues in the marine industry, and research regarding hybrid systems has already been performed. For example, Lan et al. [4] proposed a hybrid system combined with a photovoltaic (PV) generation system, a diesel generator, and batteries. Choi et al. [5] and Han et al. [6] each proposed a fuel cell–battery hybrid system for a boat. In addition, Ovrum et al. proposed a hybrid system with lithium-ion batteries (LIBs) and diesel generators for a bulk carrier [7].

However, there is not much research regarding the supercapacitor (SC) and LIB hybrid system yet, except for some research focused on small ships; Trieste et al. chose a SC as the power source for a ferry and proposed a charging strategy [8], and Bellache et al. investigated the LIB–SC hybrid system to improve the dynamic response of a boat [9]. On the contrary, many research studies have been conducted to develop the LIB–SC hybrid system for land vehicles, especially in [10–14]; these are SC–LIB hybrid systems for electric cars. These results show that the hybrid system could improve system performance by overcoming individual limitations (disadvantages) and enabling synergistic effects. In other words, LIBs, which are the most common battery types, have a high energy density; however, their power densities are low compared to that of an SC of the same size. Also, LIBs have a short life cycle compared to an SC, which has an approximately 1000× longer life cycle than LIBs (refer to Table 1).

Table 1. Comparison between lithium-ion batteries (LIBs) and supercapacitors (SCs) [15–20].

Type	Energy Density (Wh/kg)	Power Density (W/kg)	Life Cycles (cycles)	Voltage (V/cell)	Charging/Discharging Time
Lithium-ion Battery (LIB)	150~250	50~2,000	500~2,000	3.6~4.2	Minutes ~ Hours
Supercapacitor (SC)	5~10	~100,000	500,000~2,500,000	2.7~3.0	Seconds ~ Minutes

This paper proposes an SC–LIB hybrid system for a ship focused on port operations, where most emissions are caused by onboard engine-generator sets (gensets). The rest of this paper is structured as follows: in Section 2, detailed explanations of a target ship and the proposed system are presented. In Section 3, the capacity (size) of the LIB and SC is determined with the given assumed operating conditions. And in Section 4, harmful exhaust emissions at a port are calculated and compared between the conventional system and proposed one based on simulation results. Additionally, an economic study for the entire lifetime of a ship is performed in that section. Lastly, the results are reviewed along with a conclusion. The novelty of this paper is a new approach toward the eco-friendly power system of a bulk carrier using two kinds of ESSs.

2. System Description

2.1. Target Ship

In this paper, a medium-sized bulk carrier was selected as the target ship. The target ship's deadweight was about 50,000 tons, and it had five hatches. The target ship was fitted with three gensets as power sources and four electric-driven deck cranes (Figure 1). In addition to these deck cranes, windlass/mooring winches were also of the electric-driven type controlled by each motor drive. Although hydraulic-driven equipment has been used for a long time, it has many disadvantages including low efficiency, high noise and vibration, high maintenance cost, pollution risk through oil, etc. [21]. Therefore, the use of electric-driven equipment has been increasing in the marine industry.

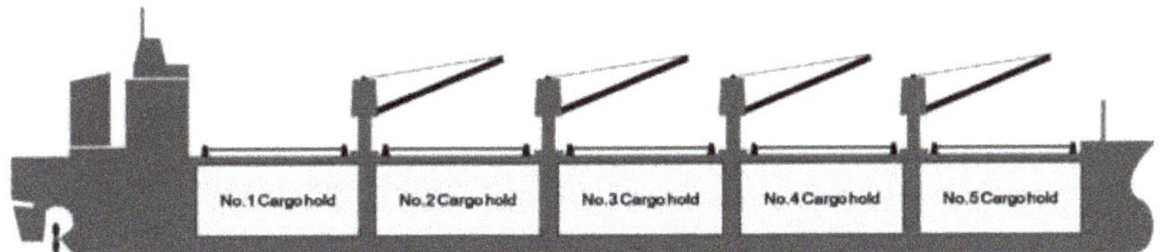

Figure 1. Typical layout of a bulk carrier with deck cranes [22].

In general, this kind of bulk carrier has four (4) operation modes, as below:

- Normal seagoing mode (at sea);
- Port in/out mode (near a port);
- Cargo loading/unloading mode (at a port);
- Harbor mode (at a port).

First, in the normal seagoing mode, the heaviest electric load is the main engine (M/E) auxiliaries and engine room auxiliaries to propel the ship, and additional electric power is required to maintain the living environment for crews at sea. In the port in/out mode, the heaviest load is the windlass/mooring winches, which are used for lowering/pulling an anchor or hauling-in/winding mooring ropes. The second heaviest load is the ballast pumps, which are used for pumping water into/out from ballast tanks in preparation for cargo loading/unloading or cargo hold cleaning. Additionally, the load on the main air compressors and M/E auxiliary blowers also increases because of the slower speed or frequent stops of a ship while approaching/departing a port.

In the cargo loading/unloading mode, the heaviest load is the onboard deck cranes used for cargo loading or unloading to the shore-side, which is a highly repetitive process. The second heaviest load is the ballast pumps, which ensures the stability of a ship even though its weight is changed during (un)loading cargo. Lastly, in the harbor mode, the majority of the load comes from the activities of crews such as from the air conditioner compressor, lighting, galley, and laundry equipment, etc.

2.2. Conventional System

The simple layout of a conventional power system is shown in Figure 2. Even though three gensets are installed as power sources, the number of gensets in operation is different depending on the power required for each operation mode. Primarily, only one genset is in operation in the normal seagoing mode with about 54.3% load factor (Table 2). The second generator is only used for the port in/out operations or the crane operations, and the last one is installed for redundancy.

Table 2. Comparison of power demands between a conventional and the proposed system.

Mode		① Normal Seagoing	② Port In/Out	③ Load/Unload	④ Harbor
Maximum demand power		380 kW	700 kW	1015 kW	250 kW
Conventional system	Power sources	G1	G1 + G2	G1 + G2	G1
	Gensets in use (load factor)	1 × 700 kW (54.3%)	2 × 700 kW (50.0%)	2 × 700 kW (71.4%)	1 × 700 kW (35.7%)
Proposed system	Power sources	G1′	G2 + LIB	AMP + SC	AMP
	Gensets in use (load factor)	1 × 500 kW (76.0%)	1 × 700 kW (71.4%)	-	-

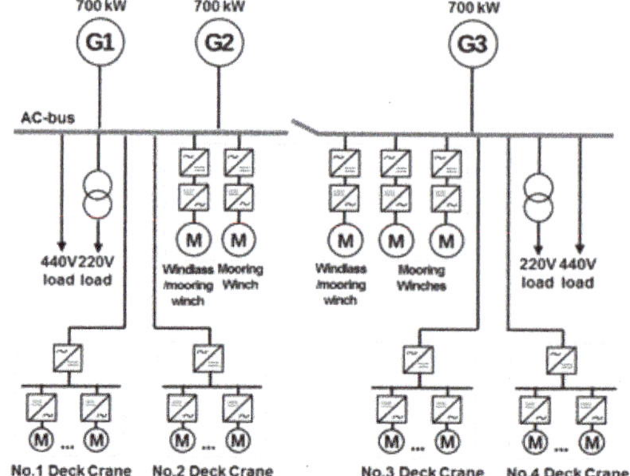

Figure 2. Layout of a conventional power system.

The steps for one voyage cycle in a conventional system are shown in Figure 3. Step 3 includes not only deck crane operation for cargo handling but also simply staying at a harbor. In this study, it was assumed that three (3) of four (4) deck cranes were in operation during cargo handling, because safety risks would be increased if all cranes were in operation simultaneously.

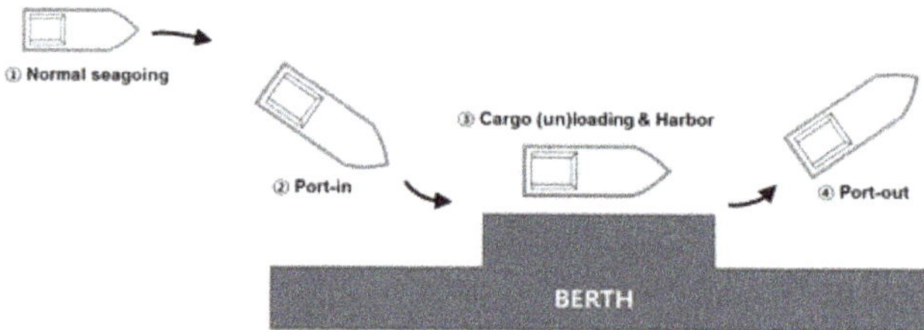

Figure 3. Docking/undocking procedure of a ship in a conventional system.

2.3. Proposed System

In the proposed system, one of the onboard gensets was replaced with two kinds of ESSs (LIB and SC). The LIB and SC were used as a power source during port operations. Also, one of the remaining gensets was downsized from 700 kW to 500 kW to obtain a higher fuel efficiency in the normal seagoing mode. The layout of the proposed power system is shown in Figure 4.

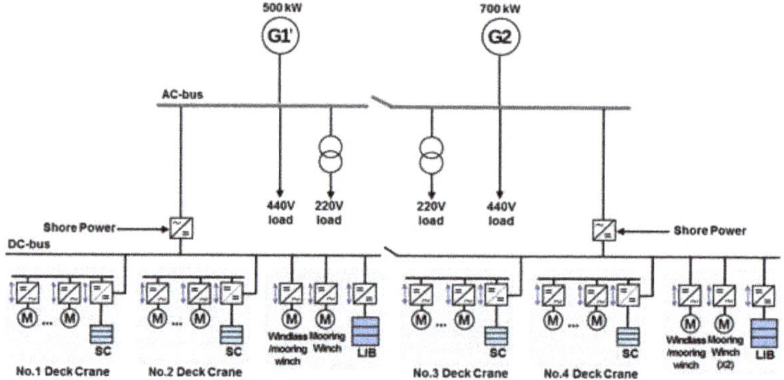

Figure 4. Layout of the proposed power system.

The reason for adopting two different ESSs was that each **has** different characteristics as an energy storage system. In the case of port in/out operations, high load demand occurred only twice (port-in and port-out). Thus, the LIB was more suitable because of its high energy density. On the other hand, the SC was more suitable for highly repetitive deck crane operations because of its long life cycle capacity and high power density. Therefore, the number of gensets in operation in each mode was changed, as shown in Table 2, and it was shown that load factors of the onboard gensets increased to above 70% even in the normal seagoing mode and port in/out mode. In the proposed system, two steps were added for one voyage cycle, as shown in Figure 5, because of the AMP connection/disconnection processes for shore power.

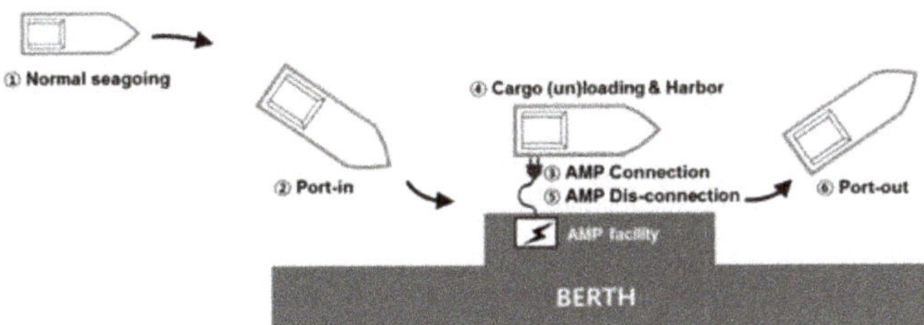

Figure 5. Docking/undocking procedure of a ship in the proposed system.

3. Proposed Hybrid Power System

The main purpose of the proposed system was to reduce harmful emissions at ports (especially in ECAs) rather than in the normal seagoing mode at sea. In the port in/out mode, in which additional power is required for a short time, the LIB was selected as an auxiliary power source that replaced a stand-by onboard genset. In the cargo loading/unloading mode, in which additional power is repeatedly required hundreds to thousands of times depending on cargo quantity, the SC and AMP were selected as the main power sources that replaced onboard gensets.

3.1. Port In/Out Mode

In this mode, the mooring and windlass winches were the heaviest loads unless bow thrusters were installed onboard. The combined windlass/mooring winch, which is used to handle both anchors and mooring ropes together, is typically installed towards the fore side of a ship, and mooring winches are installed towards the aft side. The specifications of the selected winches are described in Table 3.

Table 3. Specifications of windlass/mooring winches.

Classification		Specification
Type		**Electric-Driven Type**
Rated motor capacity	Combined windlass/mooring winch	100 kW × 2 (Forward) (one is standby)
	Mooring winch	50 kW × 3 (AFT) (one is standby)
Rated pulling force	Combined windlass/mooring winch	300 kN
	Mooring winch	150 kN
Motor drive type		Active front-end (bi-directional) type
Electric voltage		AC 440 V

The regenerated power rate when lowering an anchor was difficult to define due to various factors such as the inertia of the motor, angular speed, rotation speed, and mechanical loss, etc. [23]. Therefore, regenerated power rate was assumed to be about 50%, according to similar cases [7,24,25]. The power demands for one operation cycle were assumed, as shown in Table 4, based on empirical evidence from crews. In this study, the instantaneous peak braking and initial starting power were not taken into consideration for simulation simplification. Based on the battery manufacturer's datasheet, the specifications of the selected LIB modules are shown in Table 5.

Table 4. Expected power demands for the port in/out mode.

		Operation Mode	Time (min)	Power (kW)	Energy (kWh)
Port-in	①	Lowering (heaving up) an anchor	5	−50	−4.15
	②	Veering out and hauling in mooring ropes to the port side	10	200 [1]	33.40
		Total	**15**	**-**	**29.25**
Port-out	③	Pulling (hoisting) an anchor	10	100	16.70
	④	Winding mooring ropes to onboard rope drums	10	150 [2]	25.05
		Total	**20**	**-**	**41.75**

[1] Combined windlass/mooring winch (100 kW × 1), mooring winch (50 kW × 2). [2] Combined windlass/mooring winch (50 kW × 1), mooring winch (50 kW × 2).

Table 5. Specifications of the selected LIB module [26].

Category	Specification
Cell type	Lithium nickel manganese cobalt oxide (NMC)
Nominal voltage	88.8 V_{DC}
C-rate	Max. 1.4 C (continuous)
Stored energy	10 kWh
Capacity	112 Ah
Weight	90 kg
Size (m)	0.58 (*L*) × 0.32 (*W*) × 0.38 (*H*) (0.0705 m^3)

The minimum capacity of the LIB is calculated as shown below [27]:

$$C_{min} = (E_d \times k_a) / (V_{dc} \times k_{DoD} \times k_e) \ (Ah), \tag{1}$$

where C_{min} (Ah) is the minimum battery capacity, E_d (VAh) is the demanded energy, V_{dc} (V) is the nominal battery voltage, k_{DoD} is the battery depth of discharge (DoD), k_a is the battery aging factor, k_e is the system efficiency, and other factors are not considered. In this case, k_e was set to 0.9, k_a was set to 1.2, and k_{DoD} was set to 0.8, assuming that the operating range of the state of charge (SOC) was 10%–90%. The output voltage of the SC pack was determined to be DC 355 V, which was achieved by arranging four (4) battery modules in series. Moreover, two (2) parallel strings were required to meet the demanded energy capacity. Therefore, C_{min} was calculated to be about 196.0 Ah, and the designed battery capacity (E_{LIB_design}) was calculated to be about 76.5 kWh, incorporating a 10% safety margin (k_s) as below. The specifications of the selected LIB pack are shown in Table 6. This LIB pack was split into two sets with the same capacities, and installed at different places for safety reasons.

$$E_{LIB\_design} = C_{min} \times V_{dc} \times k_s = 196.0 \ Ah \times 355 \ V \times 1.1 \cong 76.5 \ (kWh) < 80 \ (kWh). \tag{2}$$

Table 6. Specifications of the selected LIB pack.

Category	Specification
Target terminal voltage	355 V
Configuration	2 strings × 4 modules in series
Usable energy	80 kWh
Total weight (8 modules)	720 kg
Total size (8 modules)	0.5642 m^3

3.2. Cargo Loading/Unloading Mode

In this mode, the onboard deck cranes were the heaviest loads, and they needed repetitive peak power while unloading or loading cargo. These cranes were generally required to perform three functions, namely, to hoist/lower, to luff and to slew.

- Hoisting (lowering) is bringing up (down) a crane wire while a crane jib remains in a constant position.
- Luffing is the raising or lowering of a crane jib.
- Slewing is the swinging round (or rotation) of a crane.

Among these crane operations, the biggest load is the hoist motor, which raises and lowers cargo. When lowering the cargo, the motor drive must be capable of handling the inverse power by feeding it back to the onboard main power grid. The specifications of the selected deck crane are shown in Table 7.

Table 7. Specifications of the selected deck crane.

Classification		Specifications
Crane type		Electric-driven type
Hoisting max. capacity		30 t
Max. lifting height		40 m
Crane weight		45 t
Hoisting/slewing speed		20 m/min (full load) 40 m/min (no load)
Luffing speed		10 m/min (full load)
Motor rated power	Hoisting	145 kW
	Luffing	90 kW
	Slewing	40 kW
	Grab	20 kW
Motor drive type		AFE (bi-directional) type
Electric voltage		AC 440 V

The power demand for hoisting or luffing is dependent on the weight required to carry the cargo by a crane, as shown in the below equation. The crane jip weight (m_j) is only applied to the luffing operation [28,29]:

$$P_{hoist} = (m_h \times v_h) / (6.12 \times \eta) = ((m_{load} + m_g + (m_j)) \times v_h) / (6.12 \times \eta) \text{ (kW)}, \tag{3}$$

where, m_h (t) is the hoisting weight, m_{load} (t) is the cargo load weight, m_g (t) is the grab weight, v_h (m/min) is the hoisting speed, and η is the mechanical efficiency. In this study, m_{load} was set to the maximum load of 30 t, m_g was assumed to be 5 t, m_j was 10 t, and η was 0.85. The power demand for slewing was also dependent on the weight required to turn the load by a crane, as shown below:

$$P_{slew} = (m_s \times v_h) / (6.12 \times \eta) = ((m_{load} + m_{st}) \times R_S \times v_s) / (6.12 \times \eta) \text{ (kW)}, \tag{4}$$

where, m_s (t) is the slewing weight, m_{st} (t) is the weight of the slewing structure, R_s is the resistance to slewing, v_s (m/min) is the slewing speed, and η is the mechanical efficiency. In this study, it was assumed that m_{st} was 10 t, η was 0.85 and R_s was 0.2. The regenerative power rate was assumed to be 50% according to similar cases [7,24,25] as mentioned in Section 3.1. There are ten (10) steps for a deck crane operation, and the expected power demands for one cycle are assumed, as shown in Table 8, based on the empirical evidence from crews. Then, the minimum energy of the SC pack (E_{sc_min}) was obtained through the demand energy (E_{sc_demand}), aging factor (k_a), safety factor (k_s), and the system efficiency of (k_e). If k_a was assumed to be 1.2, k_s was assumed to be 1.1, and k_e was assumed to be 0.9, The E_{sc_min} is calculated as below:

$$E_{sc\_min} = E_{sc\_demand} \times (k_a/k_e) \times k_s = 1{,}005.92 \text{ (Wh)} \times (1.2/0.9) \times 1.1 \cong 1.48 \text{ (kWh)} \cong 5{,}311 \text{ (kJ)}. \quad (5)$$

Table 8. Expected power demands for deck crane operations.

No.	Step	Time (s)	Power (kW)	Energy (Wh)
①	Lowering with no load	20	−19.2	−106.68
②	Grab (close)	10	20.0	55.56
③	Hoisting with full load	25	134.6	934.66
④	Luffing in (up)	5	86.5	120.15
⑤	Slewing to port side	15	34.6	144.18
⑥	Lowering with full load	15	−67.3	−280.44
⑦	Grab (open)	5	20.0	27.78
⑧	Hoisting with no load	10	38.4	106.68
⑨	Slewing to ship side	10	23.1	64.17
⑩	Luffing out (down)	5	−43.3	−60.14
Total		About 4 min/cycle (including overhauling time)		1005.92 Wh/cycle

In most applications, the SC pack is assembled in modules, and these are connected in series and parallel to increase both the working voltage and overall capacitance. The total capacitance of the SC pack ($C_{SC,t}$) is then evaluated as:

$$C_{SC,t} = C_{SC,module} \times (P/S) \text{ (F)}, \quad (6)$$

where P is the number of parallel strings and S is the number of series modules. The selected SC module specifications are as shown in Table 9. Then, the SC pack capacity was calculated using Equation (7).

$$E_{SC} = (1/2) \times C_{SC,t} \times V_{SC,t}{}^2 \text{ (J)}, \quad (7)$$

where $V_{SC,t}$ is the voltage of an SC pack, which is proportional to the number of SC modules. In general, the voltage variation of an SC pack is to be kept between 100% and 50% of its maximum voltage. As shown in Figure 6, the LIB offers a fairly constant discharge voltage performance throughout the spectrum of usable energy, whereas the SC voltage shows a linear and decreasing behavior from the maximum value up to 50% in general. Even if the stored energy is proportional to the product of the capacitance for the square of the voltage, the delivered power decreased in the discharging phase because the current is limited [31,32]. Thus, the available energy of the designed SC pack (E_{SC_design}) is calculated by the following equation [31,33]:

$$E_{SC\_design} = (1/2) \times C_{SC,t} \times (V_{SC,t,max}{}^2 - V_{SC,t,min}{}^2) = (1/2) \times$$
$$C_{SC,t} \times (V_{SC,t,max}{}^2 - ((1/2) \times V_{SC,t,max}))^2) = (3/8) \times C_{SC,t} \times V_{SC,t,max}{}^2 \text{ (J)}, \quad (8)$$

where $V_{SC,t,max}$ is the maximum terminal voltage, and $V_{SC,t,min}$ is the minimum terminal voltage of the SC pack. If the output voltage of the SC pack was determined to be DC 625 V, which was achieved by arranging five (5) SC modules in series, then, the SC pack needed have three (3) parallel strings to obtain the demand capacity according to the below equation:

$$E_{sc\_design} = (3/8) \times (C_{SC,module} \times (P/S)) \times V_{SC,t,max}^2 = (3/8) \times$$
$$(63 \, (F) \times (3/5)) \times (625 \, (V))^2 \cong 5{,}537 \, (kJ) \cong 1.54 \, (kWh) > 1.48 \, (kWh). \tag{9}$$

The design specifications of the SC pack are as shown in Table 10. If the SC pack was discharged, it could be charged with a C-rate of 180 within about 20.1 seconds during the luffing (10) and lowering (1) steps.

Table 9. Specifications of the selected SC module [30].

Category	Specification
Rated capacitance	63 F
Rated voltage	125 V
Max. initial equivalent DC series resistance (ESR$_{DC}$)	18 mΩ
Max. leakage current (at 25 °C)	10 mA
Number of cells	48 in series
Stored energy	140 Wh
Usable specific power	1700 W/kg
Specific energy	2.3 Wh/kg
Cycle life (at 25 °C)	1,000,000 cycles
Weight	61 kg
Size (m)	0.619 (*L*) × 0.425 (*W*) × 0.265 (*H*) (0.0697 m^3)

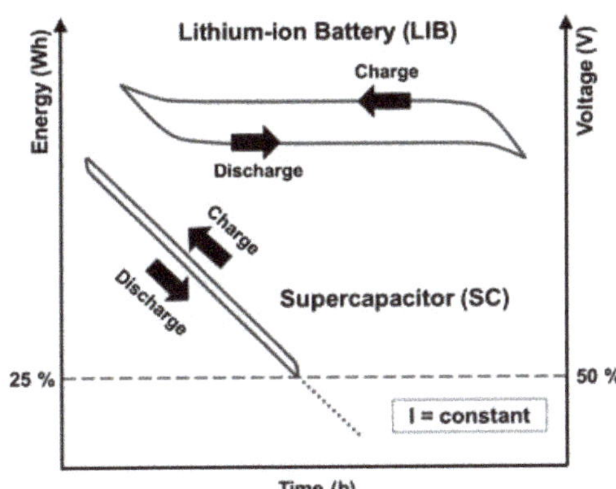

Figure 6. Comparison of charging/discharging characteristics between SC and LIB.

Table 10. Specifications of the selected SC pack.

Category	Specifications
Target terminal voltage	625 V
Configuration	3 strings × 5 modules in series
Usable energy	5537 kJ (1.54 kWh)
Total weight (15 modules)	915 kg
Total size (15 modules)	1.0455 m^3

4. Results and Discussion

4.1. Simulation Results

A simulation was conducted using MATLAB/Simulink (MathWorks, Natick, MA, USA) which is a graphics-based simulation environment to validate whether the determined capacities of the LIB and SC packs were suitable for each required power demand. Figure 7a shows the LIB pack with two parallel strings of four modules each, as indicated in Table 6. Figure 7b presents the SC pack with three parallel strings of five modules each, as specified in Table 10. Each LIB and SC pack was charged or discharged according to each required power demand, shown in Tables 4 and 8. The LIB and SC modules used in the simulation were the generic models provided in Simulink. The applied parameters were obtained from the data in Tables 5 and 9, which were based on the manufacturer's specifications [26,30], and the other parameters were assigned predetermined default values in the model (Table 11).

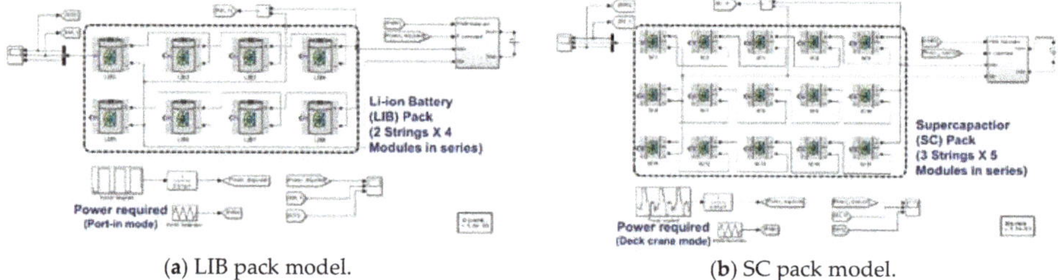

(a) LIB pack model.   (b) SC pack model.

Figure 7. Simulation models of the proposed energy storage systems (ESSs).

Table 11. Applied parameters of LIB and SC modules for the simulation.

Model	Description	Value	Unit
	Nominal voltage	88.8	V
	Rated capacity	112	Ah
Li-ion Battery (LIB)	Fully charged voltage	103.36	V
	Nominal discharge current	48.70	A
	Internal resistance	7.93	mΩ
	Rated capacitance	63	F
	Rated voltage	125	V
Supercapacitor (SC)	Surge voltage	130	V
	Initial voltage	122	V
	Equivalent DC series resistance (ESR$_{DC}$)	18	mΩ
	Leakage current	10	mA

The change in the voltage and SOC of the designed LIB pack according to the power demand is shown in Figure 8. The lowest SOC of the LIB pack was 49.3% after port-in, and 40.1% after port-out, so the capacity of the designed LIB pack was sufficient for windlass/mooring winch operation. In other words, its SOC was within the limited operating range (10%–90%). The change in the voltage and SOC of the designed SC pack during cargo handling is shown in Figure 9. The lowest SOC of each SC pack was 67.7% after the first operation cycle, so the designed SC pack capacity was sufficient for deck crane operation. In other words, its SOC was within the limited operating rage (50%–100%), and it could be recharged by shore power for a short time during steps 1 and 10, as mentioned in Section 3.2.

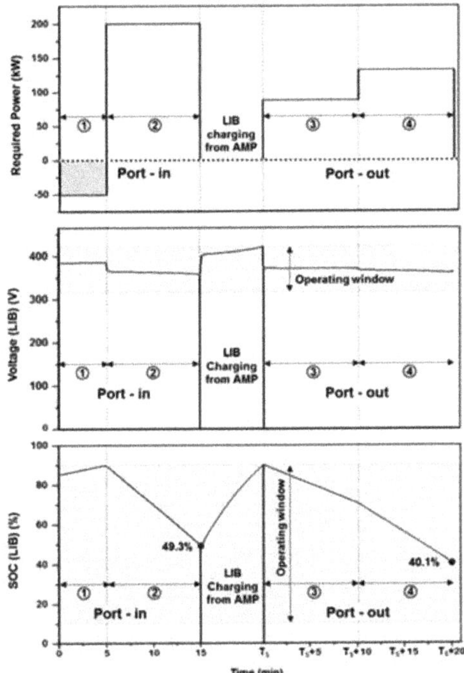

Figure 8. LIB pack voltage and state of charge (SOC) changes (port in/out mode).

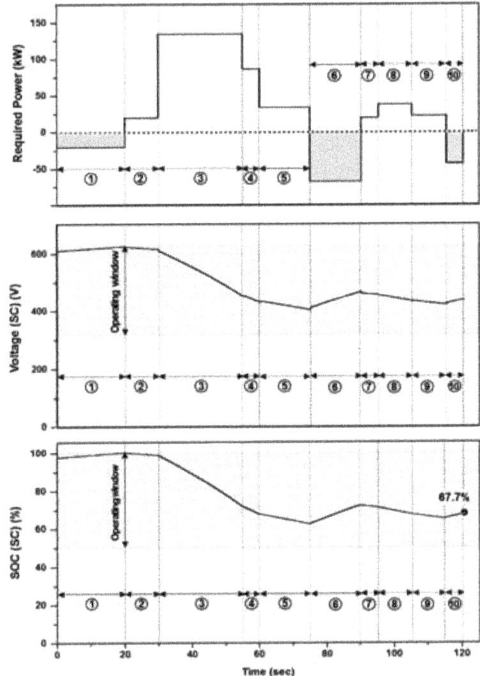

Figure 9. SC pack voltage and SOC changes (deck crane mode).

4.2. Fuel Consumption and CO₂ Emissions

The fuel consumption of a genset varies depending on the load factor, as shown in Figure 10, and the lowest fuel consumption is between 70%–85%. In this study, this graph was used to calculate the fuel consumption of the gensets. The emissions from fuels can be calculated by multiplying the fuel consumption of the onboard engine with the emission factor (E_f). This E_f varies according to the engine type (main and auxiliary engines, auxiliary boilers), engine rating, engine speed, type of fuel, etc. [34].

$$\text{Total emissions (kg)} = \text{Fuel consumption} \times E_f. \tag{10}$$

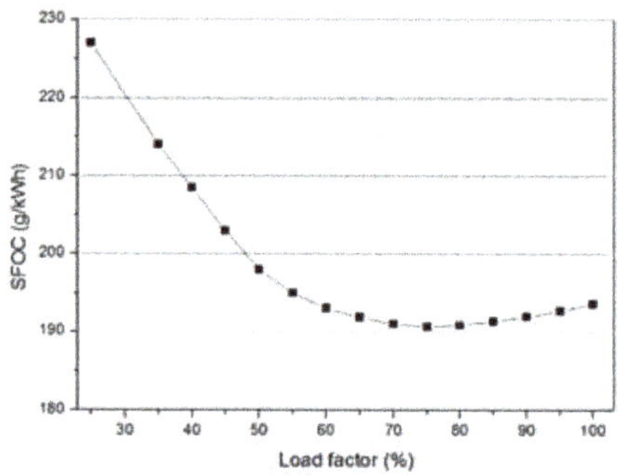

Figure 10. Example of a genset fuel consumption graph [35].

For CO_2 emissions, the E_f for each fuel type was based on IMO guidelines [36]. The E_f of HFO was 3.114 based on its lower calorific value of 40,200 kJ/kg and carbon content of 0.8493. The E_f of MGO was 3.206 based on its low calorific value of 42,700 kJ/kg and carbon content of 0.8744. And, the E_f of SO_X emissions was calculated by multiplying 0.02 with the sulfur content S (%) present in the fuel. In the case of MGO, S (%) did not exceed 0.1 %, whereas the average value of HFO was 2.7%. The E_f used for NO_X emissions was the suggested value for Tier I ships without the use of a scrubber system. These emission factors are summarized in Table 12 [37]. Based on the emission factors, the emissions from onboard gensets for the conventional power system were calculated as shown in Table 13, and those for the proposed power system were calculated as shown in Table 14.

Table 12. Emission factors for different pollutant types.

Fuel	Emission factors		
	CO_2 (g·CO_2/g·fuel)	SO_X (g·SO_X/g·fuel)	NO_X (g·NO_X/g·fuel)
Heavy Fuel Oil (HFO)	3.114	0.054	0.057
Marine Gas Oil (MGO)	3.206	0.002	0.057

Table 13. Emissions from onboard gensets for each mode (conventional system).

Mode		Electric Power Demand (kW)	Time (h)	Fuel Efficiency (g/kWh)	Fuel (kg)	Emissions (kg)		
						CO_2	SO_X	NO_X
Normal seagoing (10 days)		380	240	195	17,784.0	55,379.38	960.34	1013.69
Port in/out	Excluding winch loads	500	2 [38]	198	198.0	634.79	0.40	11.29
	Winch (Port-in)	29.25 kWh (Table 4)			5.8	18.59	0.01	0.33
	Winch (Port-out)	41.75 kWh (Table 4)			8.3	26.61	0.02	0.47
Cargo loading/unloading	Excluding crane loads	550	120 [1]	192	12,672.0	40,626.43	25.34	722.30
	Crane loads (3 cranes)	1.01 kWh × 3 each × 1800 cycle [2] (Table 8)			1047.2	3357.32	2.09	59.69
Harbor		250	48	213	2556.0	8194.54	5.11	145.69
Total			-		34,271.3	108,237.66	993.31	1953.46

[1] Assuming crane operators work in shifts of 6 h (120 h = 6 h × 20 turns) [39]. [2] Assuming each crane was operated for 15 cycles per hour (1800 cycles = 120 h × 15 cycles).

Table 14. Emissions from onboard gensets for each mode (proposed system).

Mode		Electric Power Demand (kW)	Time (h)	Fuel Efficiency (g/kWh)	Fuel (kg)	Emissions (kg)		
						CO_2	SO_X	NO_X
Ship power (AMP)	Normal seagoing (10 d) — Ship loads	380	240	191	17,419.2	54,243.39	940.64	992.89
	Normal seagoing (10 d) — LIB Charging (After port-out)	80 kWh × (85%–40.1%)		191	6.8	21.18	0.37	0.39
	Port in/out (Excluding winch loads)	500	2	192	192.0	615.55	0.38	10.94
Total			-		17,618.0	54,880.12	941.39	1004.23

Even though the ESS did not generate harmful emissions directly at a port, the emissions were generated indirectly, because it had to be recharged using the AMP; shore power was originally transferred from land-based power plants. Thus, the generated emissions from the used shore power were calculated as shown in Table 15. In this study, emission factors that generated 1 kWh of electricity were assumed to be 151 g·CO_2/kWh, 0.03 g·SO_X/kWh, and 0.16 g·NO_X/kWh based on a European electricity company [40]. This value changed depending on the country. For example, in Denmark where the dominant electricity power source is from wind power plants (about 44%, 2016) [41], the total CO_2 emission factor is 75 g/kWh, whereas the world average is 507 g/kWh [42].

Table 15. Emissions from shore charging (proposed system).

Mode		Electric Power Demand (kW)	Time (h)	Emissions (kg)		
				CO_2	SO_X	NO_X
Shore power (AMP)	Cargo loading/unloading — Excluding crane loads	550	120	9966.00	1.98	10.56
	Cargo loading/unloading — Crane loads (SC charging)	1.54 kWh × 3 each × (97.4%–67.7%) × 1800 cycle		372.95	0.07	0.40
	Harbor — LIB charging (After port-in)	80 kWh × (90%–49.3 %)		4.92	0.00	0.01
	Harbor — Harbor loads	250	48	1812.00	0.36	1.92
Total		80,502.41 kWh		12,155.86	2.42	12.88

Overall, the proposed system could reduce CO_2, SO_X, and NO_X emissions, especially in the cargo handling and harbor modes at a port (Figure 11c,d). There was about a 77% reduction for CO_2, about a 93% reduction for SO_X, and a 99% reduction for NO_X. On the contrary, the emission reduction rates for the normal seagoing mode (Figure 11a) and the port in/out mode (Figure 11b) were not high (under 10%).

In addition, the emission reduction rate varied depending on the ship's schedule. As shown in Table 16, when the cargo handling time was 60 h, the emission reduction rate was approximately 28% for CO_2, 4% for SO_X, and 35% for NO_X, but this increased to 45%, 6%, and 56% each for 180 h of long cargo handling operations. And, as shown in Table 17, when the sailing time was 20 d, the emission reduction rate was approximately 26% for CO_2, 4% for SO_X, and 32% for NO_X, but this increased to 50%, 8%, and 64% each for 5 d of short sailing time. Therefore, the proposed system is more eco-friendly if a ship has a long cargo handling time or visits many ports with a short-term sailing time.

Table 16. Comparison of emissions according to different cargo handling times.

Cargo Handling Time at a Port (h) t[1]	Conventional System (ton/yr)			Proposed System (ton/yr)			Emission Reduction (%)		
	CO_2	SO_X	NO_X	CO_2	SO_X	NO_X	CO_2	SO_X	NO_X
60	1724.91	19.59	31.25	1237.33	18.86	20.23	28.27	3.73	35.26
90	1944.83	19.73	35.16	1289.03	18.87	20.29	33.72	4.36	42.29
120	2164.75	19.87	39.07	1340.72	18.88	20.34	38.07	4.98	47.94
150	2384.67	20.00	42.98	1392.41	18.89	20.40	41.61	5.55	52.54
180	2604.59	20.14	46.89	1444.11	18.90	20.45	44.56	6.16	56.39

[1] Assuming that the ship visits 20 ports per year with a sailing time of 10 d.

Table 17. Comparison of emissions according to different sailing times.

Sailing Time for Port-To-Port (d) [1]	Conventional System (ton/yr)			Proposed System (ton/yr)			Emission Reduction (%)		
	CO_2	SO_X	NO_X	CO_2	SO_X	NO_X	CO_2	SO_X	NO_X
5 (27 ports/yr)	2174.79	13.85	39.06	1,077.69	12.78	14.06	50.45	7.73	64.00
10 (20 ports/yr)	2164.75	19.87	39.07	1,340.72	18.88	20.34	38.07	4.98	47.94
15 (16 ports/yr)	2174.84	23.58	39.36	1,506.52	22.63	24.22	30.73	4.03	38.47
20 (13 ports/yr)	2127.02	25.40	38.57	1,576.63	24.50	26.13	25.88	3.54	32.25

[1] Assuming that the ship stays at a port for 200 h with a cargo handling time of 120 h.

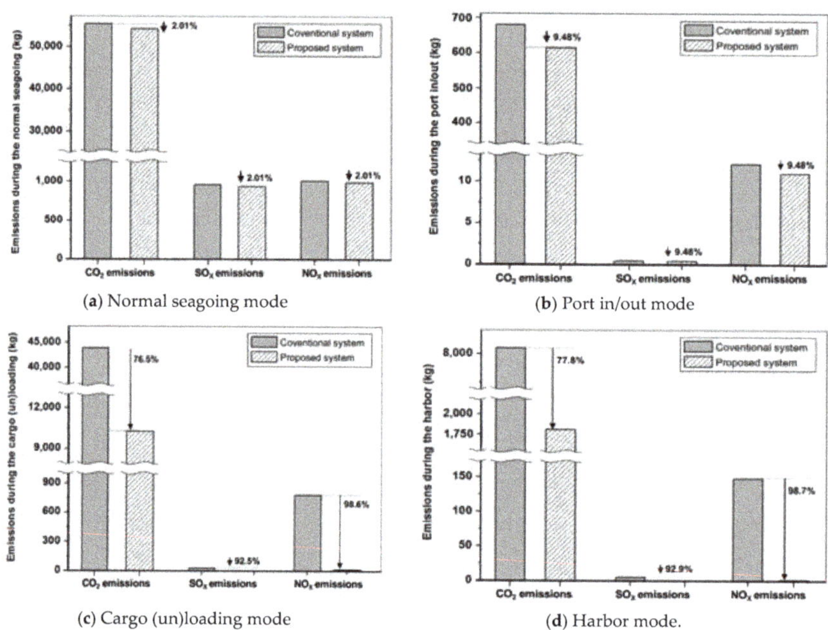

(**a**) Normal seagoing mode

(**b**) Port in/out mode

(**c**) Cargo (un)loading mode

(**d**) Harbor mode.

Figure 11. Comparison of emissions for each operation mode.

4.3. Economic Study

An economic study was conducted to compare the conventional power system and proposed one. Some assumptions were made for the study (below) since it was difficult to obtain exact data from the industry, and the data were changeable depending on the cases.

- Only the main equipment was considered;
- The bulk carrier visited 20 ports per year;
- The lifespan of the ship was 25 years.

First, the initial capital expenditure (CAPEX) is the sum of the equipment cost for the system. The cost data for the main equipment were obtained from several references [43–47]. The cost of the LIB was assumed to be 600 /kWh USD, and it was changeable according to the C-rate capacity, cell type, and cooling method, etc. The CAPEX results for the conventional and proposed systems are shown in Table 18. Secondly, the operational expenditure (OPEX) is the sum of each fixed operation & maintenance (O&M) cost, fuel cost for genset(s), and the electricity cost; only the electricity cost for the AMP was considered for the proposed system (Table 19).

The fixed O&M cost data of each main equipment were obtained from several references [48–50]. In this study, the variable O&M costs, which included cooling water or consumable materials used in maintenance, were assumed negligible because they comprised a relatively small portion in general power systems [51–53]. It was also assumed that it was necessary to switch the onboard fuel from HFO to MGO in port areas to meet environmental regulations for the conventional power system. The total savings during N years is the sum of the yearly savings (S_{year}), taking into account the interest rate (i) for the capital, as below [54]:

$$\text{Total savings} = \sum_{n=1}^{N} \frac{S_{year}}{(1+i)^n}. \tag{11}$$

In this study, i was set to 5%, and the annual inflation rates for the fixed O&M cost (a), fuel oil cost (b), and electricity cost (c) were set to 2% each. The replacement of the LIB and SC packs was considered with a replacement cost rate of 80% [55] of the initial cost. And the lifespan was assumed to be about 10 years for the LIB [56] and 15 years for the SC [17].

Table 18. Comparison of the capital expenditure (CAPEX) for each system.

Type	Conventional Power System			Proposed Power System		
	Equipment	Cost	No.	Equipment	Cost	No.
	Generator (700 kW)	149,800 USD	3	Generator (500 kW)	107,000 USD	1
	Crane converter (300 kW)	90,000 USD	4	Generator (700 kW)	149,800 USD	1
Main	-	-	-	LIB (40 kWh)	24,000 USD	2
equipment	-	-	-	SC (1.54 kWh)	15,400 USD	4
cost	-	-	-	Converter of LIB (40 kW)	12,000 USD	2
	-	-	-	Converter of SC (280 kW)	84,000 USD	4
	-	-	-	AMP converter (750 kW)	225,000 USD	2
	-	-	-	Cable system for AMP	1,300 USD	1
Total cost	809,400 USD			1,177,700 USD		

Thus, the payback period was obtained by solving for n when the initial investment cost was equal to the sum of the yearly savings. Payback occurred at around 5.8 years where the curve passed through the zero of the y-axis in the case of the below assumptions:

- The electricity cost for the AMP was 9.2 cents/kWh;
- The HFO cost was 400/t **USD**, and the MGO cost was 630/t **USD**.
- The LIB cost was 600/kWh **USD**, and the SC cost was 10,000/kWh **USD**.

However, the payback period increased to around 10 years or deceased to around 4 years according to the electricity and fuel costs, as shown in Figure 12a,b; these were more critical variables compared to the LIB or SC cost during the lifetime of a ship, as shown in Figure 12c,d. For a 25-year lifespan of a ship, the total savings would be about 0.78 million USD, and the difference was greatly dependent on the fuel and electricity cost, as shown in Figure 13. Even though electricity cost was additionally included for the proposed system, it could be economically beneficial because of the fuel savings (up to 60.2%) compared with the conventional one.

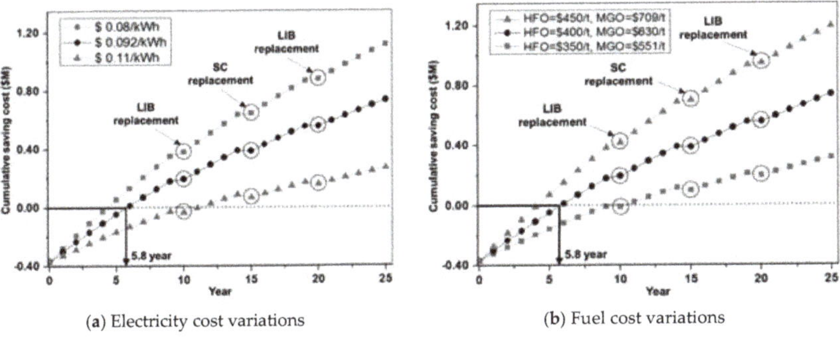

(a) Electricity cost variations

(b) Fuel cost variations

Figure 12. *Cont.*

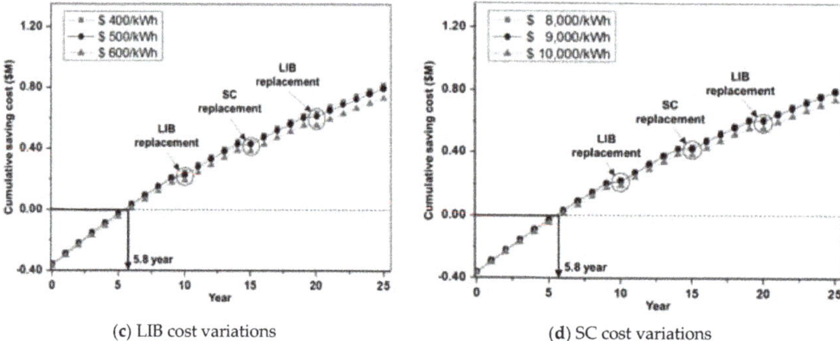

(**c**) LIB cost variations (**d**) SC cost variations

Figure 12. Cumulative saving costs for the lifespan of a ship depending on each variation.

Table 19. Comparison of the operational expenditure (OPEX) for each system per year.

Type	Conventional Power System			Proposed Power System		
	Equipment	Unit cost (/kW/yr)	No.	Equipment	Unit cost (/kW/yr)	No.
Fixed O&M cost /year	Generator (700 kW)	15 USD	3	Generator (500 kW)	15 USD	1
	Crane converter (300 kW)	2 USD	4	Generator (700 kW)	15 USD	1
	-	-	-	LIB (40 kWh)	3 USD	2
	-	-	-	SC (1.54 kWh)	5.55 USD	4
	-	-	-	Converter (LIB) (40 kW)	2 USD	2
	-	-	-	Converter (SC) (280 kW)	2 USD	4
	-	-	-	AMP converter (750 kW)	2 USD	2
Total cost	33,900/year USD			23,674/year USD		
Type	Fuel Consumption	No. of visited ports/yr	Unit price	Fuel Consumption	No. of visited ports/yr	Unit price
HFO	17.784 (t/port)	20	400/t USD [1]	17.426 (t/port)	20	400/t USD [1]
MGO	16.487 (t/port)	20	630/t USD [1]	0.192 (t/port)	20	630/t USD [1]
Electricity (AMP)	-	-	-	80,502.41 (kWh/port) (Table 15)	20	9.2 cents/kWh [2]
Total cost	350,008/year USD			289,952/year USD		

[1] BW380, BW0.1%S price (as of Jan. 2018) [57]. [2] At Halifax port (Canada) [58,59].

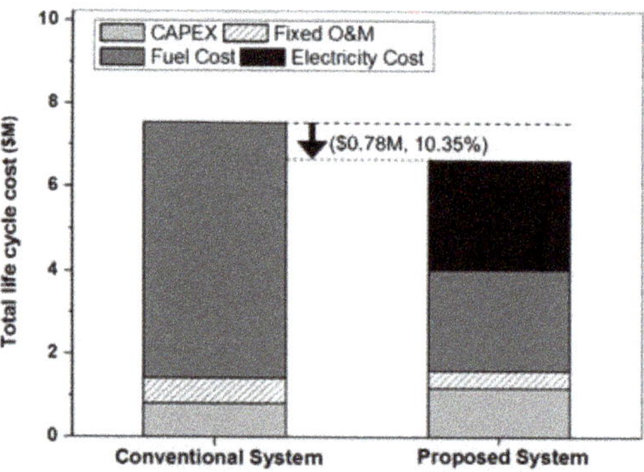

Figure 13. Total life cycle costs for conventional and proposed systems based on present values (25 years; i = 5%; a, b, c = 2%; electricity cost = 9.2 cents/kWh).

In addition, the payback period varied depending on the ship's schedule. As shown in Figure 14, when the cargo handling time was 60 h, payback occurred at around 10.2 years, but this decreased to 4.2 years for 180 h of long cargo handling operations. And, as shown in Figure 15, when the sailing time was 20 d, payback occurred at around 8.2 years, but this decreased to 4.5 years for 5 d in a short sailing time. Therefore, the proposed system is more economical if a ship has a long cargo handling time or visits many ports with a short-term sailing time.

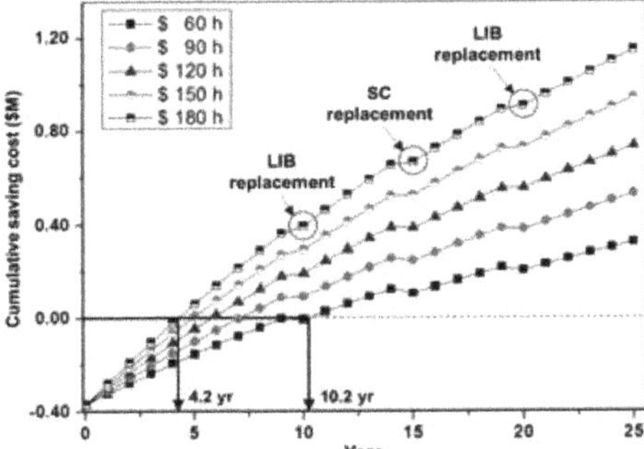

Figure 14. Cumulative saving costs for the lifespan of a ship depending on different cargo handling times (assuming that the ship visits 20 ports per year with a sailing time of 10 days).

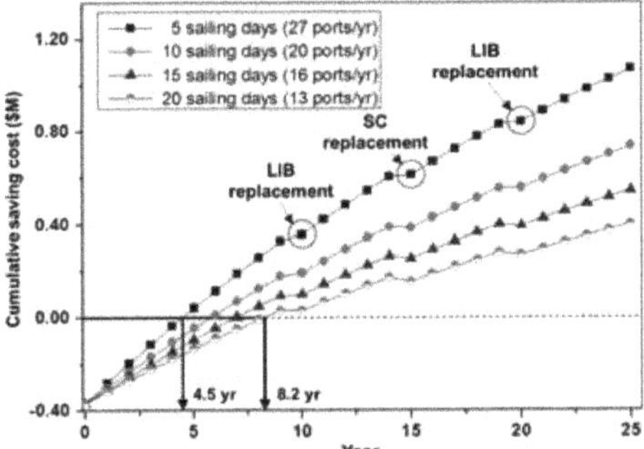

Figure 15. Cumulative saving costs for the lifespan of a ship depending on different sailing times (assuming that the ship stays at a port for 200 h with a cargo handling time of 120 h).

5. Conclusions

This paper presents a new alternative solution to reduce harmful emissions at ports, which are mostly generated from onboard gensets. The hybrid power system with two different ESS types is proposed for port operations based on a bulk carrier with deck cranes. In the target ship, the LIB is optimal for the port in/out mode, and the SC is optimal for highly repetitive deck crane operations. To verify the proposed system, the optimal sizes for the LIB and SC are determined according to the load demands, and each capacity is verified using simulations. The emission reductions are then compared with those of the conventional power system. Lastly, an economic study is performed based on the expected CAPEX and OPEX of each system.

The results show that the emission problems in port areas can be solved using this onboard hybrid power system with an AMP facility. And these environmental benefits would be increased if shore power is only generated by clean power sources such as solar power, wind power, fuel cells, etc. Moreover, the economic study shows that this proposed system will be beneficial in terms of the total lifespan of a ship. Particularly, this proposed system can be more advantageous for ships that have a long cargo handling time or visit many ports with a short-term sailing time. However, benefits are highly variable depending on the fuel oil cost for gensets and the electricity cost for the AMP. Even though this paper focused on one type of ship, the two types of ESSs (LIB and SC) could also be applied to other ship types. Therefore, this new approach to for eco-friendly ship could be helpful for many ship owners who are faced with urgent environmental regulation problems.

Author Contributions: K.K. wrote the manuscript, conceptualized, and analyzed; J.A. analyzed; K.P. and G.R. validated results; and K.C. supervised.

Appl. Sci. **2019**, *9*, 1547

Funding: This material is based upon work supported by the Ministry of Trade, Industry & Energy (MOTIE, Korea) under Industrial Core Technology Development Program. No. 10077637, 'Development of a 20MW shipboard DC optimal power system'.

Conflicts of Interest: The authors declare no conflict of interest.

References

1. Hall, D.; Pavlenko, N.; Lutsey, N. *Beyond Road Vehicles: Survey of Zero-Emission Technology Options across the Transport Sector*; ICCT (International Council on Clean Transportation): Washington, DC, USA, 18 July 2018.
2. *A Study on Demand Estimation and Implementation for AMP (Alternative Maritime Power) Installation*; KMI (Korea Maritime Institute): Busan, Korea, 2017; pp. 10–13.
3. Liu, H.; Fu, M.; Jin, X.; Shang, Y.; Shindell, D.; Faluvegi, G.; Shindell, C.; He, K. Health and Climate Impacts of Ocean-Going Vessels in East Asia. *Nat. Clim. Chang.* **2016**, *6*, 1037–1041. [CrossRef]
4. Lan, H.; Wen, S.; Hong, Y.-Y.; Yu, D.C.; Zhang, L. Optimal Sizing of Hybrid PV/Diesel/Battery in Ship Power System. *Appl. Energy* **2015**, *158*, 26–34. [CrossRef]
5. Choi, C.H.; Yu, S.; Han, I.S.; Kho, B.K.; Kang, D.G.; Lee, H.Y.; Seo, M.S.; Kong, J.W.; Kim, G.; Ahn, J.W.; et al. Development and Demonstration of PEM Fuel-Cell-Battery Hybrid System for Propulsion of Tourist Boat. *Int. J. Hydrogen Energy* **2016**, *41*, 3591–3599. [CrossRef]
6. Han, J.; Charpentier, J.-F.; Tang, T. An Energy Management System of a Fuel Cell/Battery Hybrid Boat. *Energies* **2014**, *7*, 2799–2820. [CrossRef]
7. Ovrum, E.; Bergh, T.F. Modelling Lithium-Ion Battery Hybrid Ship Crane Operation. *Appl. Energy* **2015**, *152*, 162–172. [CrossRef]
8. Trieste, S.; Hmam, S.; Olivier, J.-C.; Bourguet, S.; Loron, L. Techno-Economic Optimization of a Supercapacitor-Based Energy Storage Unit Chain: Application on the First Quick Charge Plug-in Ferry. *Appl. Energy* **2015**, *153*, 3–14. [CrossRef]
9. Bellache, K.; Camara, M.B.; Dakyo, B. Transient Power Control for Diesel-Generator Assistance in Electric Boat Applications Using Supercapacitors and Batteries. *IEEE J. Emerg. Sel. Top. Power Electron.* **2018**, *6*, 416–428. [CrossRef]
10. Camara, M.B.; Gualous, H.; Gustin, F.; Berthon, A. Design and New Control of DC/DC Converters to Share Energy Between Supercapacitors and Batteries in Hybrid Vehicles. *IEEE Trans. Veh. Technol.* **2008**, *57*, 2721–2735. [CrossRef]
11. Thounthong, P.; Raël, S.; Davat, B. Control Strategy of Fuel Cell/Supercapacitors Hybrid Power Sources for Electric Vehicle. *J. Power Sources* **2006**, *158*, 806–814. [CrossRef]
12. Kouchachvili, L.; Yaïci, W.; Entchev, E. Hybrid Battery/Supercapacitor Energy Storage System for the Electric Vehicles. *J. Power Sources* **2018**, *374*, 237–248. [CrossRef]
13. Sadoun, R.; Rizoug, N.; Bartholumeus, P.; Barbedette, B.; LeMoigne, P. Sizing of Hybrid Supply (Battery-Supercapacitor) for Electric Vehicle Taking into Account the Weight of the Additional Buck-Boost Chopper. In Proceedings of the 2012 First International Conference on Renewable Energies and Vehicular Technology, Hammamet, Tunisia, 26–28 March 2012. [CrossRef]
14. Mih, B.; Svasta, P. Hybrid supercapacitor-battery electric system with low electromagnetic emissions for automotive applications. *UPB Sci. Bull.* **2013**, *75*, 277–290.
15. *Carbon-IonTM: A New, Safer & Faster Charging Category of Rechargeable Energy Storage Devises*; White Paper; ZapGo: Oxford, UK, 2017.
16. Santoso, S. *Standard Handbook for Electrical Engineers, Seventeenth Edition*, 17th ed.; McGraw-Hill Education: New York, NY, USA, 2017.
17. Gidwani, M.; Bhagwani, A.; Rohra, N. Supercapacitors: The near Future of Batteries. *Int. J. Eng. Inven.* **2014**, *4*, 22–27.
18. Lin, Y.-L.; Kyung, C.-M.; Yasuura, H.; Liu, Y. (Eds.) *Smart Sensors and Systems*; Springer International Publishing: Cham, Switzerland, 2015. [CrossRef]
19. Zakeri, B.; Syri, S. Electrical Energy Storage Systems A Comparative Life Cycle Cost Analysis. *Renew. Sustain. Energy Rev.* **2015**, *42*, 569–596.
20. Rodriguez-Martinez, L.M.; Omar, N. (Eds.) *Emerging Nanotechnologies in Rechargeable Energy Storage Systems*; Elsevier: Amsterdam, The Netherlands, 2017.

21. Lamas Pardo, M.; Carral Couce, L.; Castro-Santos, L.; Carral Couce, J.C. A review of the drive options for offshore anchor handling winches. *Brodogradnja* **2017**, *68*, 119–134. [CrossRef]

22. Isbester, C.J. *Bulk Carrier Practice Polymath*; ExC FNI; The Nautical Institute: London, UK, 1993; p. 206.

23. *ABB Drives—Technical Guide No. 8, Electrical Braking*; ABB: Zurich, Switzerland, 2011; p. 13.

24. *STARVERT IRU. Technical Report*; LS Industrial Systems Co., Ltd., LS Industrial Systems: Anyang, Korea, 2009; p. 2.

25. *Proposal Paper—High Performance Bi-Directional Inverters*; Hyundai Motor Industry: Seoul, Korea, 2014.

26. System Specifications for the PBES Power & Energy Systems (Module: Energy 100). *PBES (Plan B Energy Storage)*, 30 May 2017.

27. IEEE. *Standard 485: IEEE Recommended Practice for Sizing Lead-Acid Batteries for Stationary Applications*; IEEE: Piscataway, NJ, USA, 2010.

28. Agrawal, K.C. *Industrial Power Engineering Handbook*; Elsevier: Amsterdam, The Netherlands, 2001.

29. Yaskawa Electric Corporation. *Permanent-Magnet Synchronous Motor SS7-Series Eco PM Motor*; Yaskawa Electric Corporation: Fukuoka, Japan, 2016; p. 23.

30. Maxwell Technologies. 125V Heavy Transportation Module (BMOD0063 P125 B08). Available online: http://www.maxwell.com/products/ultracapacitors/downloadsTechnicalDatasheet (accessed on 8 September 2018).

31. Ciccarelli, F. Energy Management and Control Strategies for the Use of Supercapacitors Storage Technologies in Urban Railway Traction Systems. Ph.D. Thesis, University of Naples, Napoli, Italy, 2014.

32. Reveles-Miranda, M.; Flota-Bañuelos, M.; Chan-Puc, F.; Pacheco-Catalán, D. Experimental Evaluation of a Switching Matrix Applied in a Bank of Supercapacitors. *Energies* **2017**, *10*, 2077. [CrossRef]

33. Barrero, R.; Tackoen, X. *New Technologies (Supercapacitors) for Energy—Storage and Energy Recuperation for a Higher Energy Efficiency of the Brussels Public Transportation Company Vehicles*; Prospective Research for Brussels (Intermediate Report); Université Libre de V (ULB): Bruxelles, Belgium, 2008; p. 48.

34. de Germán, M.R.; Carlos, M.G.J.; Enrique, M.A.; Sergi, S.M. Evaluation Air Emission Inventories and Indicators Form Ferry Vessels at Ports. In Proceedings of the 12th International Conference on Marine Navigation and Safety of Sea Transportation (TransNav 2017), Gdynia, Poland, 21–23 June 2017.

35. MAN Diesel & Turbo. *Marine GenSets from MAN Diesel & Turbo*; MAN Diesel & Turbo: Gothenburg, Sweden, 2014.

36. *MEPC.1/Circ.866 Annex: 2014 Guidelines on the Method of the Attained Energy Efficiency Design Index (EEDI) for New Ships*; IMO: London, UK, 30 January 2017.

37. Zis, T.; Angeloudis, P.; Bell, M.G.H.; Psaraftis, H.N. Payback Period for Emissions Abatement Alternatives: Role of Regulation and Fuel Prices. *Transp. Res. Rec. J. Transp. Res. Board* **2016**, *2549*, 37–44. [CrossRef]

38. *Commercial Marine Vessels Review*; EPA (United States Environmental Protection Agency): Washington, DC, USA, 2015.

39. Mrs, M. *Safety Investigation into a Stevedore Fatality on Board the Maltese Registered Bulk Carrier—TARSUS. Marine Safety Investigation Report 13*; MSIU (Marine Safety Investigation Unit): Floriana, Malta, 2013; p. 45.

40. *ESG (Environment, Social and Governance) Performance Report*; Ørsted: Fredericia, Denmark, 2017; p. 14.

41. *Environmental Report for Danish Electricity and CHP for 2016 Status Year (Doc. No.: 16/19207-5)*; Energinet: Erritsø, Denmark, 2017.

42. *IEEJ Outlook 2018—Prospects and Challenges until 2050. The 427th Forum on Research Work*; IEEJ (The Institute of Energy Economics): Tokyo, Japan, 2017; p. 85.

43. Hekkenberg, R.G. A Building Cost Estimation Method for Inland Ships. In Proceedings of the 7th European Inland Waterway Navigation Conference (EIWN), Budapest, Hungary, 10–12 September 2014.

44. How Does a Supercapacitor Work? Battery University. Available online: http://batteryuniversity.com/learn/article/whats_the_role_of_the_supercapacitor (accessed on 15 August 2018).

45. IDC (Industrial Development Corporation). *Release of the US Trade and Development Agency Sponsored Energy Storage for South Africa Study—Energy Storage Could Become a Future Industry in South Africa*; IDC (Industrial Development Corporation): Sandton, South Africa, 2017; p. 45.

46. Klein, P. *Energy Storage Use Cases for Enhancing Grid Flexibility*; SAPVIA Networking Event; Council for Scientific and Industrial Research (CSIR): Petroria, South Africa, 2017; p. 20.

47. Winkel, R.; Weddige, U.; Johnsen, D.; Hoen, V.; Papaefthymiou, G. *Potential for Shore Side Electricity in Europe. Final Report*; Navigant Consulting (formerly Ecofys): Chicago, IL, USA, 2015; p. 21.

48. *Lazard's Levelized Cost of Energy Analysis—Version 8.0*; Lazard: New York, NY, USA, 2014.

49. Viswanathan, V.; Kintner-Meyer, M.; Balducci, P.; Jin, C. *National Assessment of Energy Storage for Grid Balancing and Arbitrage Phase II*; Pacific Northwest National Laboratory (PNNL): Washington, DC, USA, 2013; p. 78.

50. Spatarua, C.; Kokb, Y.C.; Barretta, M. Physical Energy Storage Employed Worldwide. *Energy Procedia* **2014**, *62*, 452–461. [CrossRef]

51. Sim, S. *Electric Utility Resource Planning: Economics, Reliability, and Decision-Making*; CRC Press: Boca Raton, FL, USA, 2011.

52. Sabol, S. *Case Studies in Mechanical Engineering: Decision Making, Thermodynamics, Fluid Mechanics and Heat Transfer*; John Wiley & Sons: Hoboken, NJ, USA, 2016.

53. Kelp, O.; Lenton, R.; Choudhuri, G. *Fuel and Technology Cost Review—Final Report*; Acil Allen Consulting: Melbourne, Australia, 2014.

54. Mukund, R.P. *Shipboard Electrical Power Systems*; CRC Press: Boca Raton, FL, USA, 2017.

55. Rodrigues, A.; Machado, D.; Dentinho, T. Electrical Energy Storage Systems Feasibility; the Case of Terceira Island. *Sustainability* **2017**, *9*, 1276. [CrossRef]

56. *Battery Energy Storage Study for the 2017 IRP (PacifiCorp)*; DNV GL KEMA: Arnhem, The Netherlands, 2016; p. 15.

57. Fuel Prices (BW380). Available online: http://www.bunkerworld.com/prices/bunkerworldindex/bw380 (accessed on 30 January 2018).

58. *Tariffs—Shore Power Tariff*; Nova Scotia Power: Halifax, NS, Canada, 2017; p. 36.

59. *Application for Approval of Shore Power Rate*; BC Hydro: Vancouver, BC, Canada, 2015; pp. 1–11.

MDPI

St. Alban-Anlage 66

4052 Basel

Switzerland

Tel. +41 61 683 77 34

Fax +41 61 302 89 18

www.mdpi.com

Applied Sciences Editorial Office

E-mail: applsci@mdpi.com

www.mdpi.com/journal/applsci